P9-CKY-597

The DIAMOND DIGGERS
South Africa 1866 to the 1970's

Also by Ivor Herbert

BOOKS

Over our Dead Bodies!

The Way To The Top

The Winter Kings

The Queen Mother's Horses

Arkle: the Story of a Champion

Point-to-Point

Eastern Windows

THEATRE

Night of the Blue Demands

CINEMA

The Great St. Trinian's Train Robbery

The Ferret

Numerous documentaries

TELEVISION

Hyperion

IVOR HERBERT

The Diamond Diggers

South Africa 1866 to the 1970's

Tom Stacey

First published in 1972 by
Tom Stacey Ltd. 28–29 Maiden Lane,
London WC2E 7JP

ISBN 0 85468 151 5

Printed in Great Britain by
Compton Press, Compton Chamberlayne, Salisbury

For
MY FRIENDS
and
FOREBEARS
in
South Africa

◆

ACKNOWLEDGEMENT Any book about the diamonds of South Africa is so largely concerned with the growth of 'De Beers' and 'Anglo American', that it would be impossible to write any authentic account without their help.

People in De Beers Consolidated Mines Ltd, and Anglo American Corporation of South Africa Ltd., in Kimberley, London and Johannesburg, went so far out of their way to help me and were so friendly about doing so, that I must note my special gratitude to them here, mentioning particularly Mr. Basil Humphreys of Kimberley who showed me round, instructed me and then, from hospital, corrected my proofs.

I.H.

Bradenham,
Buckinghamshire.

Contents

Illustrations and maps

Between pages 92 and 93

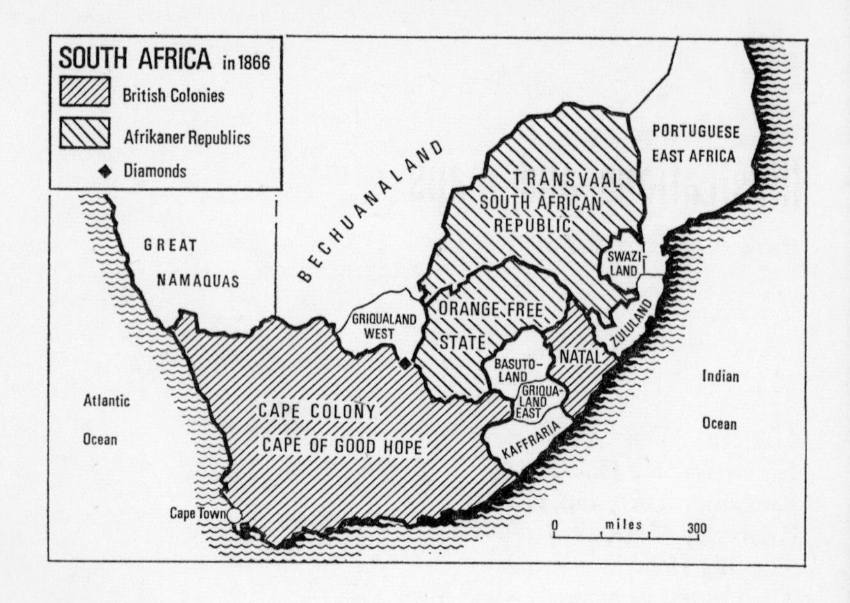
SOUTH AFRICA in 1866
British Colonies
Afrikaner Republics
Diamonds
BECHUANALAND
PORTUGUESE
EAST AFRICA
TRANSVAAL
SOUTH AFRICAN
REPUBLIC
SWAZI-
LAND
GREAT
NAMAQUAS
GRIQUALAND
WEST
ORANGE FREE
STATE
ZULULAND
BASUTO-
LAND
NATAL
GRIQUA-
LAND
EAST
KAFFRARIA
Indian
Ocean
Atlantic
Ocean
CAPE COLONY
CAPE OF GOOD HOPE
Cape Town
0 miles 300

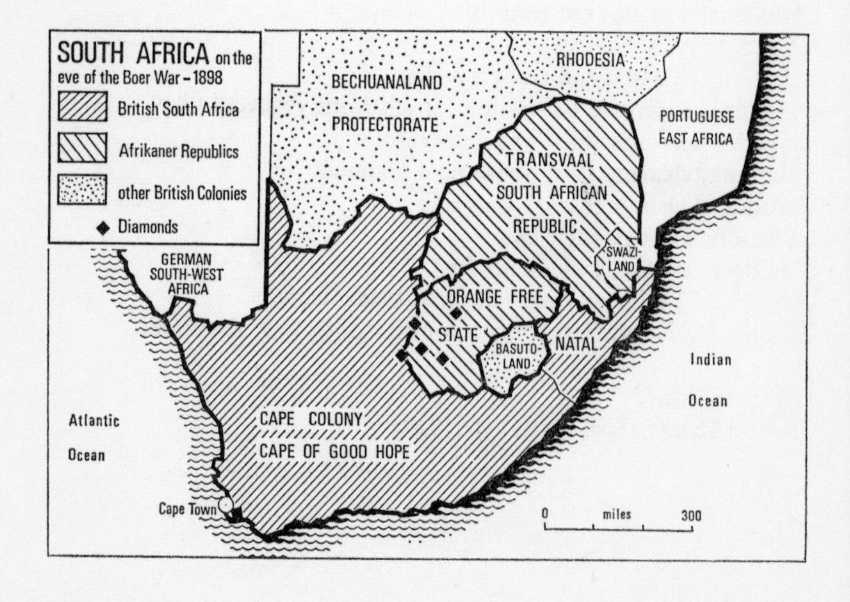
SOUTH AFRICA on the
eve of the Boer War – 1898
British South Africa
Afrikaner Republics
other British Colonies
Diamonds
BECHUANALAND
PROTECTORATE
RHODESIA
PORTUGUESE
EAST AFRICA
TRANSVAAL
SOUTH AFRICAN
REPUBLIC
SWAZI-
LAND
GERMAN
SOUTH-WEST
AFRICA
ORANGE FREE
STATE
BASUTO-
LAND
NATAL
Indian
Ocean
Atlantic
Ocean
CAPE COLONY
CAPE OF GOOD HOPE
Cape Town
0 miles 300

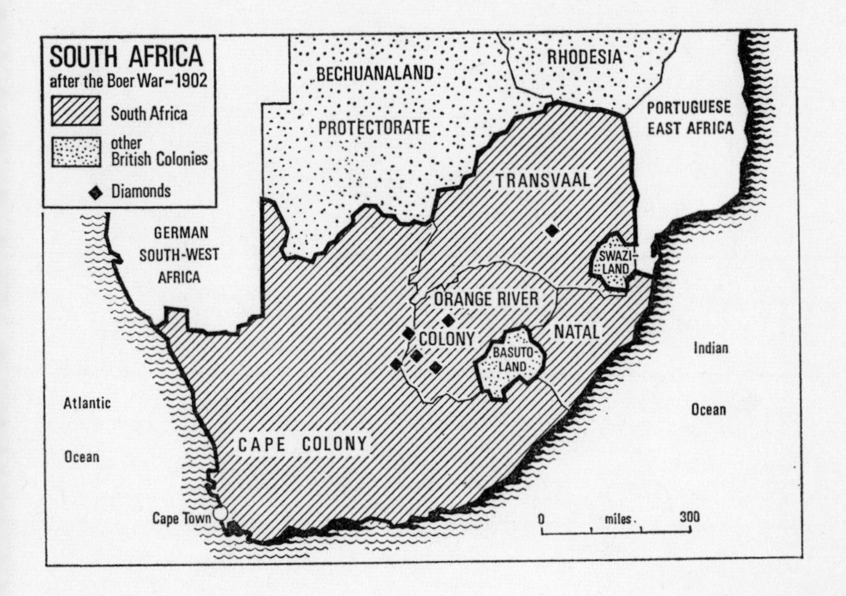
SOUTH AFRICA
after the Boer War–1902
South Africa
other British Colonies
Diamonds
BECHUANALAND
PROTECTORATE
RHODESIA
PORTUGUESE EAST AFRICA
TRANSVAAL
GERMAN SOUTH-WEST AFRICA
SWAZI-LAND
ORANGE RIVER
COLONY
NATAL
BASUTO-LAND
Indian
Ocean
Atlantic
Ocean
CAPE COLONY
Cape Town
0 miles 300

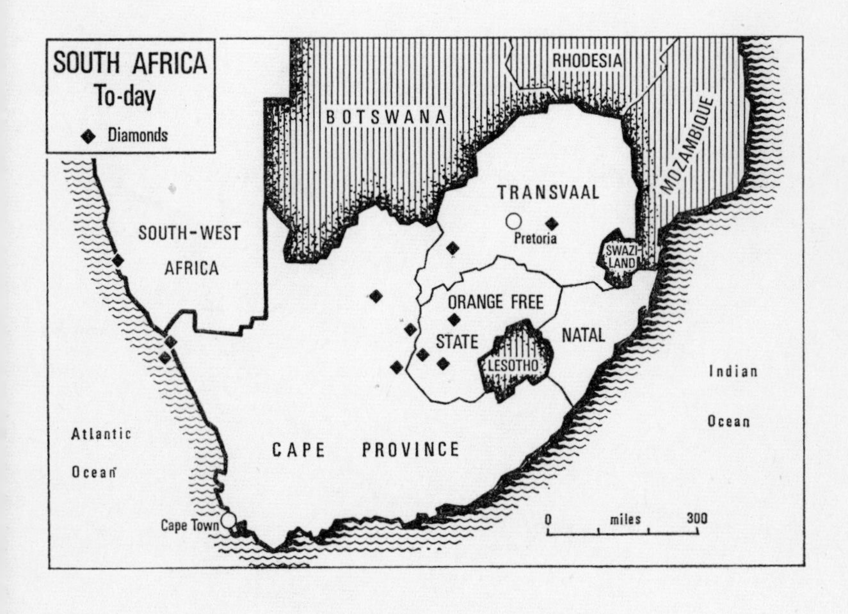
SOUTH AFRICA
To-day
Diamonds
BOTSWANA
RHODESIA
MOZAMBIQUE
TRANSVAAL
Pretoria
SOUTH-WEST
AFRICA
SWAZI-LAND
ORANGE FREE
STATE
NATAL
LESOTHO
Indian
Ocean
Atlantic
Ocean
CAPE PROVINCE
Cape Town
0 miles 300

CHAPTER ◆ ONE

Baby Jacobs' peculiar Pebble

The farmers of southern Africa in the middle of the last century owned great tracts of veld, but they were not vastly richer than the Africans they employed. Their children, owning few toys, played with pebbles. Some stones of pretty colours shone as the bright sun smote the bare land.

The children were Boers, a word derived from the Dutch word for farmer, and they chattered to each other in the new tongue of Afrikaans, a language crudely built on old Dutch foundations with additions of corrupted German, French and English. The language is now spoken naturally by 2 million South Africans and officially by a few hundred thousand more and Afrikaners tell you proudly that Afrikaans is thefirst new language in the world in the last 1,000 years. Its maintenance and encouragement against all odds reflect the character of those early Boers whose children were playing with unrecognized garnets and diamonds on the dirt floors of small raw-brick shacks while their fathers struggled to keep alive and free.

Nobody knew about diamonds in South Africa in the middle of the nineteenth century in that dubiously frontiered hinterland which is now the heart of one of the world's richest countries: the Republic of South Africa. In the last thirty years of the last century the discovery first of diamonds then of gold turned a poor agricultural land into a young industrial giant.

The road to the country's riches began in the spring of 1866 on the banks of the Orange River. On a large farm called 'de Kalk' lived the Boer Daniel Jacobus Jacobs. The farms were all large in terms of acres or of morgen, the term used in South Africa for measuring land (a morgen is 100 yards × 100 yards). There was no shortage. New settlers had only to gallop out from their oxwaggons on their ponies to claim the area they could encircle in a day. But there was little agricultural substance in the land. Jacobs ranched cattle and sheep, grew maize, kept chickens. He was worth something in terms of capital, but his income, particularly in a dry year, was painfully low.

The nearest settlement was called Hopetown, an optimistic name for a huddle of shacks on a dusty track. The nearest towns were at least 400 miles away on the sea: Cape Town to the south-west, Port Elizabeth to the south-east and the beginnings of Durban far away to the west, were the only places larger than villages in all southern Africa and they were all in the hands of the British. The Boers who had generally been there first had trekked inland into the wilds to try to farm in peace. The modern city of Johannesburg lay unconceived below the rolling veld.

Of the Boer family Jacobs it was their mother, named in Afrikaans Mevrou Jacobs, who first drew attention to a particular pebble being tossed around by her children. It so sparkled in the sunlight that she showed it to her neighbour Schalk van Niekerk the next time he rode over.

The claimant for the honour of finding the stone was her son Erasmus Stephanus Jacobs, born on 23rd October, 1851. He was an old and somewhat vague gentleman speaking little English when he was questioned in 1930 by the local historian in Kimberley, Mr. A. J. Beet.

He had been out to clear a water pipe for his father some way from the homestead, two hundred yards from the banks of the Orange River. He sat down for a rest under a tree, then suddenly noticed something flashing. 'It blinked,' he told Mr. Beet, snapping his fingers in and out. So he brought it home and gave it to his baby sister to play with.

The children had long since cast the gleaming stone aside and Mevrou Jacobs and van Niekerk had to root about in the dirt of the farmyard to find it again. Under its glossy surface it had a peculiar glow.

'I'd like to buy it,' said van Niekerk, thinking it might possibly be a sort of topaz.

Mevrou Jacobs laughed. 'Nonsense. It's a present.'

The diamond left in the neighbour's leather pocket. Van Niekerk was the local Divisional Councillor and ranched some stock himself, which was to prove fortunate three years later. He lived in a house on the Jacobs' land, was generally respected and knew a lot of people at a time when few Boers knew anyone except their next-door neighbour ten miles' ride away across the veld.

Whether van Niekerk guessed what he had in his pocket when he rode into Hopetown, no one has been able to establish. When great things are done it is tempting to claim the glory in old age and to say: 'It wasn't just luck.'

Van Niekerk handed over the stone to bearded Jack O'Reilly, a pedlar and crack big-game shot who was based on Colesberg in the Karoo, away to the south-east. But no one knows whether van Niekerk sold him the stone, or whether he asked him, on a commission basis, to show it to the Government mineralogist in Grahamstown further on to the south. It was described then in the rough as being 'nearly white and resembling a piece of white alum, except that it was extremely heavy'. And there were of course those dazzling spots where the outer cloudiness had been polished off.

Some sources say O'Reilly paid van Niekerk a few pounds for the stone, but all agree that the Irishman knew he had a diamond. He cut his name on a window pane with it, and had no doubts. In his travels, grinding hundreds of miles in his cart round remote farmsteads, dealing in necessities and gossip, he must have heard of a missionary, the Rev. John Campbell, who had made his second visit to South Africa in 1820.

Campbell crossed the Transvaal, camped by the Orange River and sat down to make up his diary: 'One of our people on first coming to this river collected many kinds of bright stones. He is now looking them over and throwing many of them away, having better knowledge. Age and experience discover many things to be trifles which in youthful days were highly esteemed.'

But Campbell did not entirely believe they were all 'trifles'. When he got home he made a map of his journey which was finally published in 1852 for anyone to read in a missionary journal. And all across that area of the Orange River valley Campbell wrote 'Here be diamonds.' Nobody can have believed him, for it was fourteen years before van Niekerk passed the Jacobs' pebble on to O'Reilly.

This blundering across the face of facts is typical of the diamond saga in South Africa. Fantastic riches were certainly lying deep in the earth, like the hidden gold, platinum and uranium which no one could find without digging. But alluvial diamonds on dry African river banks had been lying around on the surface for centuries, ridden over by missionaries and travelling pedlars, trampled into the sparse tufts of grass by thirsty cattle, picked up by ostriches, played with by black children (and later by white) and kicked aside by bare black feet or by the farmers' leather home-made shoes.

O'Reilly returned to his base at Colesberg (now a small town in the Karoo) and showed the stone to the Civil Commissioner, Mr. Lorenzo Boyes. The Commissioner was excited enough to agree that it probably *was* a diamond, and O'Reilly journeyed on to visit the Grahamstown mineralogist, Dr. W. G. Atherstone.

Atherstone tested the stone. The Roman Catholic Bishop of Grahamstown cut his name on glass with it. The doctor and the bishop blunted files on it. It was definitely a diamond, said Dr. Atherstone. But what was it worth? Nobody knew. There was no other local product in Africa as a criterion. Jewellery was unknown on the farms. Expert jewellers did not crowd the lonely veld.

Among the poor farmers and young British officers there was no market for large diamonds. 'About £500,' hazarded Dr. Atherstone, and at that price lucky O'Reilly sold it to the Governor of Cape Colony, Sir Philip Wodehouse. Sir Philip sent it to Paris to be shown at the 1867 Exhibition, and the stone of $21\frac{1}{4}$ carats and perfectly white ('blue-white', as dealers say, trying to express that totally colourless clarity) attracted considerable interest. But only as a freak. It occurred to no one that the discovery of this beautiful gem heralded the emergence of more, let alone the eruption of shoals of diamonds which were going to make a cluster of millionaires, a pack of bankrupts and a colossal industry.

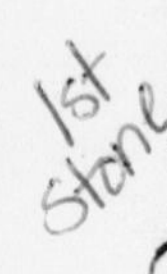

The Jacobs' stone was named 'Eureka' and was cut down to $10\frac{3}{4}$ carats (a sub-average weight loss). Later it was set in a bracelet surrounded by South African stones mostly from the Kimberley area, in 1946. This bracelet was bought at Christie's London Salesroom by Mr. Peter Locan, who resold it to De Beers Consolidated Mines in 1966. De Beers repatriated it to South Africa, exhibited it in Johannesburg to mark South Africa's Diamond Centenary and then presented it to the South African Government.

'Eureka' had come full circle. A hundred years earlier O'Reilly was enjoying a special fame. The local farmers around Hopetown and along the Orange River kept their sharp eyes under their big black hats turned well to the ground as they rode across the veld, or plodded around their cattle. Lorenzo Boyes had spurred off from Colesburg to find van Niekerk as soon as he had heard the mineralogist's report. He thrust north-west down the Orange River, discovered where the diamond had been found, and spent two desperate weeks raking over the gravel along the sides of the river. He found nothing.

Other local Boer children hoped to ape the Jacobs family and dug about in the mixed shingle for the prettiest pebbles. But there had only been one stone to show them what to look for, and few of them had caught a glimpse of it. No one in the area was capable of recognizing a raw rough stone without an expert to demonstrate its virtues.

A few insignificant stones were picked up to the north along the Vaal River, and groups of farmers rushed down to its banks to grovel through them for any bright stone. A few had sold their land to finance their prospecting. But only a handful of stones emerged. Money was short, prospects were low, and after a year interest waned. It was not as if the Jacobs family had made money from their find. Years later Erasmus Jacobs was in such straits that the citizens of the new city of Kimberley (which had not been thought of when 'Eureka' was found) passed the hat round to raise £30 for his old age. Nor had van Niekerk enriched himself from his part in the first diamond deal.

The Boers shrugged and went back to their farming. They were the most rural and parochial of people and some of their Afrikaner descendants today have inherited these traits which make them difficult for sophisticated foreigners to understand. These old Boer settlers did not care about Paris or Europe or an English Governor way down in Cape Town. They weren't ambitious. They were self-taught, barely literate, zealously religious, and proud of their freedom. They were introspective and stubborn. Their farms were their worlds. They wanted peace there from the officious British who had pushed them out of the sweet lands in the Southern Cape to this harsh bare veld where the sun scorched every day except when the heavens blackened with torrential storms.

Summer blazed from August to February and then as autumn came in hot and golden in the March of 1869, the diamond business exploded once again. Away from the Orange River and up on the high farmland of Zandfontein, an African shepherd picked up a huge diamond. The shepherd was a Griqua, a race of mixed Bushman, Hottentot and Boer blood who inhabited that part of Africa, calling it

Griqualand. The shepherd, nicknamed 'Swartbooi' (black boy) by the local Boers, took the huge stone to the witch doctor in his local kraal. The witch doctor sagely suggested that he approach van Niekerk, the only man around who knew about these glittering stones.

Boldly Swartbooi asked van Niekerk 400 sovereigns for his diamond. Van Niekerk possessed nothing like that money. But he knew he was looking at a diamond four times larger than the stone O'Reilly had sold in the Cape. If the Griqua was bold, van Niekerk was heroic. He appraised his stock and offered almost its entirety to the flabbergasted seller: 500 sheep, 10 oxen and a horse. When an ox or two made a native a rich man, and a horse might comprise his life's savings, van Niekerk's offer convinced Swartbooi that the Boer was insane. Accepting the offer instantly before the white man should return to normal Swartbooi went off to get prodigiously drunk for days.

Van Niekerk rode furiously into Hopetown and weighed his stone on some scales: $83\frac{1}{2}$ carats! The little place buzzed with the staggering news. And in moments he had sold it to a merchant for a sum between £10,000 and £11,500. The price seemed incredible when huge farms passed for mere hundreds of pounds. He returned rejoicing, rich enough to work no more for the rest of his life.

But the story and the stone sped onwards and upwards, for this gem was to become the marvellous 'Star of South Africa'. The merchant sold it on to Louis Hond, a diamond cutter who fashioned the rough stone into an oval three-sided brilliant of $47\frac{3}{4}$ carats of superlative lustre and of the first water. Hond sold it for £30,000 to the Countess of Dudley (after whom the marvellous stone is sometimes called), who had it mounted in a magnificent tiara surrounded by 95 smaller diamonds.

The 'Star of South Africa' set the southern end of the continent pulsing and the word rushed around the globe. There were indeed great diamonds to be found and great fortunes to be made. The 'Eureka' had been no fluke. All over southern Africa the treks began. Farmers from Cape Town 650 miles to the south-west and from Natal as far to the east, harnessed their teams of oxen. Their long covered waggons started to roll and creak north and west across the empty veld towards Hopetown and the confluence of the westward-flowing Vaal and Orange Rivers. To many the Vaal looked the best bet and hordes of diggers swarmed down the Vaal like wasps along a trail of jam and buzzed and probed along its two tributaries, the Modder and the Harts.

Panting behind came prospectors from overseas sailing into Table Bay and Durban from the old diamond diggings in India and Brazil; gold-diggers from Australia and North America with experience, and a host of Europeans with none. Frothing with hope and greed and frantic zeal they swarmed into southern Africa and up into the emptiness of Griqualand. The rush had started.

e Making of a Diamond

Diamond had been a magic name and a priceless jewel from antiquity. The Dravidians, who discovered diamonds in India 2,700 years ago, also gave us the word *carat*, deriving from the Carob tree's seeds which are of remarkably uniform weight. Seeds were the traditional weight measures for light objects: in medieval England one barley grain was the lowest unit. Antique pearl dealers made use of the regular uniformity of the dry seeds of the locust-pod tree (*Ceratoria Siliqua*) to measure their wares – another derivation of carat from 'cerat'. Even on modern precision scales the difference between these seeds averages no more than 1/1000th of an ounce. One carat weighs 200 milligrams, five carats making one gram, (defined by the International Committee on Weights and Measures in Paris in 1907). First used in England by the Normans, one carat represents 4 1/142nd of an ounce.

The English word 'diamond' derives from the Greek *adamas* (*'Αδαμας*) meaning 'the invincible', which the Greeks applied to any particularly hard metal and stone. The Romans knew diamonds well. Manilius referred to them in AD 16 and Pliny, writing of six varieties in AD 100 opined, 'It is the most valuable of gems, known only to kings.' The stones came westwards into the Roman Empire from India via Persia, Egypt and Greece. Later Roman writers described 'Indian rivers yielding up diamonds from their sands.'

Because it was the hardest, most brilliant and permanent of jewels, diamond often had magical qualities ascribed to it. It was thought to be a protector in battle and was therefore worn on the breastplates, helmets and sword-handles of rich princes and great captains. It was believed to be an aphrodisiac, to render poison harmless and to avert madness. The diamond was regarded in medieval Italy as a peace-maker, the *pietra della reconciliazione*. Today, promoted by powerful advertising, and in the Orient as well as in the western world, a diamond is believed to be essential for every engagement ring. And for those females who prefer to be engaged briefly, frequently and professionally, a diamond remains 'a girl's best friend'.

What is remarkable is that the coveted gem is *chemically* identical with graphite or blacklead. Yet it possesses utterly different *physical* characteristics.

Dr. Frederick Raal, D.Phil., Rhodes Scholar, and Research Manager of De Beers Diamond Research Laboratory outside Johannesburg explained: 'Diamond is regarded as the prototype of crystals which show covalent bonding – the whole crystal can be regarded as one giant molecule. Diamond, like graphite, consists of the element carbon. Each carbon atom is tetrahedrally bonded to four neighbouring atoms and provides one of the two electrons common to each bond.

'As a consequence, carbon atoms are held together by strong covalent bonding, which makes diamond extremely hard and gives it a high refractive index. Both these properties of course contribute to its value as a gem.

'For a non-metal, diamond has an exceptionally high thermal conductivity – at room temperature it is about four times as good a conductor of heat as copper. This explains why diamonds always feel cold to the touch. When heated above 500°C diamond is attacked by oxygen, and carbon dioxide is formed. In a vacuum it reverts to graphite at temperatures in excess of 1,600°C. Its melting point is estimated to be in the region of 3,600°C.'

The exact origin of diamonds remains mysterious. In simple terms they were formed millions of years ago by a combination of immeasurable heat and pressure. Subsequently they were volcanically erupted on to, or near to the earth's surface. So far no clear pattern has emerged suggesting where diamonds *should* be found, though there are several clues to likely sources.

The two simple properties of a diamond which make it attractive to man are its beauty and its hardness. But the full loveliness of diamonds could not be exposed to the western world until the early fourteenth century. Only then were stones first faceted and properly polished. Ludwig Berghem of Bruges was first reputed to have invented the art of polishing diamond in 1476, but recent evidence shows that diamonds were cut in Europe in the early 1300's.

The oldest form of a polished precious stone was called a cabochon. The type persists: it entails smoothing off the rough edges to produce a lenticular shape. These were probably the basis of the first spectacles (as Dr. S. Tolansky suggests in *The History and Use of Diamond*).

But if diamond is the hardest substance, what could polish it without wearing itself away? 'Diamond-cut-diamond' means exactly what it says. Yet, when a diamond is physically at least five times harder than the next known substance (some scales register ten times) how can it be cracked quite easily?

Dr. Raal in his cool office in the complex of research buildings at the Crown Mine outside Johannesburg explained again:

'Diamond is composed of carbon atoms in a highly symmetrical arrangement. Its hardness is due to the fact that atoms must be removed from the surface of the diamond *against* these strong bonding forces. But these forces are *directional*: a diamond can only be polished in certain directions. The diamond polisher looks carefully for the *grain* of the stone before he starts. The existence of grain means a diamond can be easily split or cleaved despite its hardness when sheer force is applied at right angles to the carbon bonds.'

Before the brittleness of diamonds was understood, thousands of fabulous gem stones were destroyed by ignorant hammering. Pliny (who was asphyxiated in AD 79 while reporting Vesuvius' eruption) declared 'a diamond can be subdued by soaking in fresh warm goat's blood. Then, if hit with such force as to break both hammer and anvil, it will yield. When it yields it falls into such small pieces they can scarcely be seen.'

The magical properties of fresh warm goat's blood have not continued down the centuries, but this was probably an early alchemist's camouflage for his discovery that a sharp blow at the right point and angle would shatter a diamond. In the unenlightened Middle Ages, diamonds squeezed in iron vices pressed themselves into the gripping metal without damaging themselves.

In the early days of South Africa's diamond rush, diggers often tested their diamonds by bashing them with a hammer. By ill chance they often hit the wrong spot. Bad luck, they would think, brushing the pieces aside, not a diamond after all. And unscrupulous dealers would play the same trick, waiting only till the disappointed digger had turned his back to shuffle sadly off, before sweeping up the chips to sell to a merchant.

CHAPTER ◆ THREE

The lure of Famed Diamonds

Because all the early Indian diamonds were found in river beds the Indians believed they grew there in the gravel and were refreshed each spring after winter's high waters – which in reality had merely washed up and cleaned off more new stones. This belief that all diamonds were alluvial was carried into Africa via Europe and misled the early prospectors along the Vaal and Orange rivers.

Indian diamonds were found in ancient sandstones, in beds often less than one foot thick, which also contained pebbles of jasper and quartzite – two other clues Africa's first diamond diggers sought for. Indian mining had been concentrated in five groups on the eastern side of the Deccan plateau and the general centre of the industry and the main diamond market, was the fortress of Golconda. When the French merchant Tavernier toured the area in the middle of the seventeenth century he found 60,000 primitive workers digging shallow pits in gravel which was first sluiced with water and then sifted by hand. These then became the methods used along the South African rivers.

In the next century the Indian fields gradually became worked out, although a huge stone of $210\frac{1}{2}$ carats was dug up at Hira Khund in 1809 and one of $67\frac{3}{8}$ carats at Wajra Karur in 1881. But the industry was already ailing when the discovery in 1725 of new fields in Brazil at Minas Geraes dealt it a death blow. Till then India had produced all the famous stones of antiquity. The glamour which shines out of diamonds started with great stones like the 'Hope', the 'Koh-I-Noor', the 'Orloff', the 'Great Mogul' and the twin 'Arcot' diamonds.

The drama of these huge historic Indian stones glowed in the minds of the African prospectors, as they wondered if they might find another fabulous blue stone like the 'Hope', the final $44\frac{1}{2}$-carat product of the 112½-carat stone brought by Tavernier from Golconda in 1642. On his return from the East, Tavernier sold it to Louis XIV. After cutting and polishing it weighed 67½ carats and became the showpiece of the French Regalia, but during the French Revolution it was stolen with all the rest from the Garde Meuble. All trace vanished.

Then in 1830 a diamond of a similar strange colour but weighing only 44½ carats, suddenly appeared on the London market. It was bought by Henry Hope, a rich banker, for £18,000 in the belief that it was the 'Blue Tavernier' smuggled out of France and recut to escape detection. In 1908 the Sultan of Turkey acquired it for £80,000. Three years later the stone was sold at Cartier's in Paris to Mrs. Edward B. McLean of Washington. At the 1949 sale of her collection it was bought by Harry Winston, the New York diamond merchant who presented the 'Hope' to the Smithsonian Institution in Washington in 1958 at a valuation of $1,000,000.

The history of the 'Koh-I-Noor' also started in the Golconda mines. Discovered before 1300 it reputedly weighed 800 carats in the rough. Two centuries later it was acquired by Sultan Baber, the founder of the Mogul dynasty. The stone stayed at Delhi till 1739 when Nadir Shah of Persia invaded north-west India, plundered Delhi and seized the 'Koh-I-Noor' and several other large diamonds with it. He is said to have shouted out 'Koh-i-noor!' ('Mountain of light') when he first gazed astonished on the stone.

On his death his empire broke up and the diamond finally returned to India, coming into the hands of Ranjit Singh, 'The Lion of the Punjab', in 1833. But in 1849 it was taken by the East India Company as partial indemnity after the Sikh Wars in the Punjab. The Company presented it to Queen Victoria in 1850 by the hand of Lord Dalhousie, at an Imperial levee marking the 250th anniversary of the founding of the East India Company by Queen Elizabeth I.

When the 'Koh-I-Noor' was displayed at the Great Exhibition held in London's Hyde Park in 1851 experts were disappointed by its lack of brilliance and Queen Victoria decided to have it recut. It was reduced to a 108⅞-carat shallow brilliant in an operation taking 38 days and carried out in London by Voorsanger, a diamond cutter specially brought over from Amsterdam. The Queen was pleased and wore the diamond frequently as a brooch. After her death it was set in the centre of the front cross of the State Crown of Queen Alexandra and then of Queen Mary. Finally it was set in the crown made in 1937 for the coronation of Queen Elizabeth the Queen Mother, and is now with the Crown Jewels in the Tower of London.

Two other legendary diamond names beckoned on the new South African prospectors: the 'Orloff' and the 'Arcot'.

The 'Orloff', a 199$\frac{1}{8}$-carat rosecut stone of exceptional purity was mounted in the Imperial Sceptre of Russia and is now in the Diamond Treasury in Moscow. Legend says the stone was stolen by a French soldier from the eye of a Brahmin idol in Madras. The Frenchman sold it to an English sea captain who resold it in London to a dealer.

In 1774 the diamond appeared in Amsterdam and was bought there by Prince Gregory Orloff, one of Catherine the Great's favourites and a leader in the conspiracy which brought down her husband Peter III. The Empress had first contemplated marrying Orloff but then he fell from favour. He bought the diamond to give to Catherine in a desperate bid for reinstatement. The Empress coolly accepted the diamond and had it mounted in the Imperial Sceptre, but rejected Orloff.

The two 'Arcot' stones, pear-shaped diamonds weighing a total of 57$\frac{1}{3}$-carats, and now privately owned in Texas, were presented by the Nawab of Arcot in 1777 to Queen Charlotte, wife of George III. Queen Charlotte bequeathed them to her four remaining daughters, directing that they should be sold and the proceeds divided. Her executors sold the stones to Rundell, Bridge and Co., Jewellers and Silversmiths to the Crown. On the death of John Bridge the firm was sold and on 20th July 1837 the 'Arcot' diamonds were bought by the first Marquis of Westminster for £11,000 as a birthday present for his wife. (He also bought the 'Nassak' diamond for £7,200 and another for £3,500.)

The 'Arcot' diamonds were mounted in the Westminster tiara, together with the 3,500 brilliants and 1,421 smaller diamonds. In 1959 the executors of the 2nd Duke of Westminster sold this at Sotheby's for £110,000 to Harry Winston of New York, at a then world record price for a piece of jewellery.

CHAPTER ◆ FOUR

The Rush to the Rivers

The Indian diamond fields were almost worked out as the South African diamond rush began, so among the prospectors heading inland to the Vaal and Orange Rivers came men with Indian experience.

The old diamond industry in Brazil (a monopoly since 1772 of the Portuguese Government which then ruled the country) had been given a fillip in 1844 when rich finds were made of carbonado – black, cryptocrystalline masses of interlocked crystals. Though Brazilian diamond recovery was still active among the alluvial gravels of rivers, the news from Africa came hot and strong, and a number of Brazilian diggers sailed east across the south Atlantic to join the throngs converging on the bare land in the centre of southern Africa.

Further north, five ships were fitting out in Boston in 1871 to bring American diggers across to the new fields of Africa.

Apart from India and Brazil, diamonds have been found in few other places. Australia has produced some, not many, but of good quality, Borneo has fielded a few small stones, and so has the USA. Several diamonds have been unearthed in California and Arkansas, but the largest American stone weighs only 40 carats. In Russia, good diamonds were found in 1829 in the Ural Mountains and in 1879 in Siberia and there are extensive diamondiferous areas in northern Yakuts – exactly how large is not yet known to the western world.

Diamond also exists in space: tiny stones have been extracted from fallen meteors, and space exploration may well reveal more.

The South Africa of 1869 on which the diamond diggers suddenly descended was a far cry from today's Republic. There was in fact no country called South Africa then at all, but a hotch-potch of races engaged in elbowing each other out of the way, a situation which was going to be aggravated by any sudden wealth. The largest part of southern Africa was occupied by Cape Colony, which covered the area of the first diamond discoveries. The colony was governed in Cape Town by an elected local majority under the aegis of London. Though the Cape of Good Hope had been discovered by Portuguese Bartholomew Diaz in 1488, Portugal never established a settlement there, and it was nearly two centuries before the Dutch East India Company settled at Table Bay in 1652, supported by a small garrison under van Riebeeck to guard the port.

By purchase and force the early Dutch colonists acquired land from the indigenes, whom they called Hottentots. In 1686 a flock of French Huguenots driven out of France by the anti-Protestant Edict of Nantes set sail for southern Africa. They settled in the Cape, bequeathing the French place-names of the present day, and later merged with the Dutch. The Hottentots consequently lost more of their land to the new European settlers and started to work for them as farm-labourers. A pattern of apartheid began to form nearly three centuries ago.

Others of the Cape's indigenous inhabitants were the Bushmen who were widely spread over the inland plains right up to the Orange River. As the Dutch colonists spread north and east the Bushmen, deprived of the game on their former hunting grounds, fell back but raided the Dutchmen's sheep and cattle. The Dutch East India Company resolved to exterminate them, formed commandos, hunted them down like foxes and in six years had practically eradicated them as a race. Today only a few scattered remnants linger on, lurking in isolated areas.

The Company was ruled from Holland. It imposed restrictions on the Dutch settlers, keeping all trade to itself, closing the colony to free immigration, and decreeing, ruling and punishing with a harsh hand. After a nasty taste of colonial discipline, the settlers started to trek away from Cape Town. They developed that dislike of any interference from Europe which was to heighten over the centuries and remains paramount today. Their attitude was hardly mollified when in 1795 Britain seized Cape Town to preclude it from French hands. During the uneasy peace in Europe in 1803 the British left the Colony, but in 1806, after war had again erupted, they recaptured Cape Town and the colony was formally ceded to Britain in 1814.

During the nineteenth century there were frequent wars among th

Kaffirs, a dark, Bantu-speaking people incorporating different tribes. They were far more numerous and powerful than the Hottentots, who, like the Bushmen, were a yellow-skinned race.

This diverse Bantu race had been loosely called 'Kaffirs' or 'Caffres' by the early European travellers – a disparaging name meaning 'unbeliever'. They are known to have lived in southern Africa for more than 1,000 years because their metal-smelting habits have been accurately carbon-dated to the Iron Age. The British, in an effort to broaden their Cape Town base and to colonize southern Africa, sent out first the '1820 Settlers' to Port Elizabeth, and later the 'Natal Settlers'. These were bands of worthy independents, adventurous in spirit and fretting in Britain, who were encouraged by London to seek their fortunes in Britain's two new colonies. Their presence in the most fertile parts of the country naturally displeased the Bantu-speaking natives. The cruel Kaffir Wars ensued, resulting in the far stricter enforcement of British rule.

The Boers too had reason to resent British authority, and decided to move out *en masse*. The Great Trek started in 1836 and over the next six years 7,000 trekkers set out for the wilds to form their own republics beyond the Orange and Vaal Rivers. They also settled briefly in Natal, still a dependency of Cape Colony and the home of a few British families. As the Boers approached the area where diamonds lurked in the gravel, trouble was flaring with Basutos, Bushmen and Griquas. The Cape Government on principle always backed the natives against the Boers so they now formally recognized Griqualand West.

By 1847 the British, although accepting the new Boer Republic of the Transvaal, had extended Cape Colony northwards to the Orange River, and were engaged in asserting their authority over the Boers beyond it. There, with the abandonment of the British Orange Colony, the Boers had declared their own Orange Free State and were now embroiled in a tedious struggle with the Basutos whom the British formally protected.

In this confusion, the first of 5,000 diamond diggers rushing down on the Vaal river knew little and cared less about whose sovereignty they might be infringing. The area, like most of southern Africa, certainly seemed poor enough: it took 8 acres of the bald Karroo to feed one sheep. Now a few square feet of diamondiferous ground was going to feed dozens of families.

Diamonds were first found in 1869 at Klipdrift, later renamed Barkly West. The area had been settled since the dawn of African Man: artefacts of the Early, Middle and Later Stone Ages abound there. Nearby the famous Glaceated Pavements, dating back 200 million years to South Africa's Ice Age carry some excellent rock-paintings believed to be 2,000 years old. Although bigoted supporters of white superiority decline to admit it, there were certainly some dark-skinned inhabitants of southern Africa thousands of years before the Dutch first called.

More recently Bushmen had roamed the area, followed by the Korana branch of the Hottentots. The latter had virtually died out from assault, famine and intermarriage with the encroaching Griquas who had started to move into the territory a century before the discovery of diamonds. The Griquas chose as their centre the confluence of the Vaal and Orange rivers, so that in 1869 one of their chiefs, Nicolaas Waterboer, promptly claimed sovereignty over the new diamond diggings. The Boer Republic of the Orange Free State immediately countered that the diggings lay in its territory.

They were not quarrelling over the empty veld, for by 1870 there were no less than 10,000 people in a vast squalid camp along the river diggings, as considerable and regular discoveries put the word about. The method of mining was as simple as it had been in India and Brazil. Gravel and sand were washed in wooden cradles screened with perforated metal and the resulting concentrate was picked over by hand on tables. A Diggers' Committee administered the area, and limited claims to squares of 30 feet each with access to the river bank.

Opposite the first diggings at Klipdrift a new centre was begun across the water at Pniel, a German Mission Station, and from there spread out 50 miles along each bank. There was barely room for all the prospectors, and those who had not yet struck it rich swarmed out further into the countryside like hounds searching for a new scent.

So it was that towards the close of 1870 and in the new year of 1871 diamonds were found again far away from the Vaal River. The finds were on the farmland of Jagersfontein near Fauresmith, about 100 miles from the Vaal River, and at Dutoitspan and Bultfontein, some 80 miles to the north of the Orange River.

A second great prospecting rush now broke like a wave of pent-up water. Farmers who had sold their land to go prospecting along the

Vaal riverbed now cursed themselves for giving away what might well be diamondiferous soil. Then in July 1871 diamonds were unearthed in what is now the Kimberley mine, and the search became frantic.

So far, to the bewildered frustration of the first diggers, diamonds had cropped up without pattern or reason. The old Indian alluvial theory had already been discounted by the rich finds on high farm land. But the prospectors for lack of new guidance (and they are still without any firm knowledge today) clung to old faiths. Because at first it seemed the new finds were all in 'pans', those slight depressions common in South Africa, the early prospectors imagined they were still alluvial in a way. Digging below the surface soil soon revealed diamonds in a layer of yellow clay about 50 feet deep. Below the clay lay the 'blue ground' now famous as the matrix of South African diamonds, but originally supposed to be barren bedrock. When it was revealed that the yellow clay was merely decomposed 'blue ground' the searchers dug desperately deeper and deeper into the earth.

They also dug wider, but the 'blue ground' appeared only as circular patches or 'pipes' shooting up to the surface from the bowels of the earth in vertical or near vertical shafts. Within an area no more than 3 miles across, five of these 'pipes' were struck near Kimberley. These five were Dutoitspan, Bultfontein, Wesselton, De Beers and Colesberg Kopje, as the Kimberley open mine – the world-famous 'Big Hole' – was first called.

CHAPTER ◆ FIVE

Diamonds on every Farm

There was soon a 10-acre working area at what is now Kimberley and a 23-acre field at Dutoitspan. Within weeks, onto what had been empty land of thorn trees, poor scrub and thin grazing restlessly cropped by a few springbok, cattle and sheep, there now rushed an uncontrolled army of 50,000 men. Suddenly in the wilderness there was spread out a city of twenty thousand tents, waggons, oxen, camp fires, dogs and piles of bare provisions, all scattered about as greed, guile, experience or exhaustion compelled. Within the chaos, the smell and dirt surpassed the 'stews' of the Middle Ages and the scant food was filthy and desperately expensive.

The land, which had previously just maintained a few farmers and Africans, was now swamped by an invasion it could do nothing to support. Every single article required by 50,000 diggers had to be humped nearly 700 miles up country from Cape Town, mostly by creaking ox-waggon from the far distant railhead. Everything, even water, stood at a ludicrous premium. One cabbage cost 10s. Most of the diggers took their only green vegetable from the weeds on the veld. A digger had to start rich or very soon become successful to afford to live in the squalid encampments at all, and the roughness and danger of the life kept the rich away.

Like much else in the South African diamond story, the second great rush had simple grass roots and almost comic beginnings. The farm Jagersfontein lay in territory occupied by the Griquas. It had been owned by the Boer family Visser since before the arrival of the earliest Voortrekkers from the Cape, so they were among the first white settlers to reach that part of Africa. The hereditary Griqua chief, Adam Kok III, made a weird deal with the Visser family. Whenever Kok and his wife visited Jagersfontein, the resident Visser wife had to pay tribute by stripping off all the clothes she was wearing at the time and handing them to Kok's wife. Fortunately this custom had fallen into disuse in 1870 when there was only an old widow Visser on the farm. She was helped by a foreman, Jaap de Klerk, the bearer of a famous old Cape Dutch name. De Klerk found some garnets in a dried-up stream, knew they were often pointers to diamonds and started to dig. Within weeks he unearthed a diamond of 50 carats, rode into Fauresmith and sold it to a courier for a few pounds. It was the courier selling it to a merchant at a glittering profit who launched the rush on Jagersfontein.

On the Vaal the diggers were paying farmers 10s a month for their rights. Old widow Visser charged £2. Soon she had over a thousand diggers scrabbling over her land, paying her tribute at the rate of £2,000 a month, which was more than the whole place was worth for sale the day before de Klerk struck lucky among the garnets.

The next three farms to yield diamonds lay 80 miles north-west of Jagersfontein. These, the combined Dutoitspan and Dorstfontein farm, then Bultfontein and finally Vooruitzicht* had originally comprised one large flattish holding of about 60 square miles. It was not enormous by the standards of those early Trekboers, who acquired new lands by the simple expedient of riding around them, but it had been split up.

Its three parts were now about to start their upward leaps in value. The financiers had not been much excited by the Vaal River operations. Monthly fees for digging rights had been paid by individuals (somewhat ephemerally) to the farmer claiming ownership. Now, however, diamonds began to smell like big business and the money boys moved in.

The diggers were already in action on Dutoitspan, owned by a Boer named van Wyck, when a syndicate bought it over their heads for £2,600. But nothing except the ownership changed. The syndicate were pleased to let the self-elected and generally respected Diggers' Committee police the place and they continued to let out the Dutoitspan rights on a monthly basis.

Then in the South African spring, a rival syndicate bought Bultfontein farm next door for £2,000 from Cornelius de Plooy. Next door again lay the farm called Vooruitzicht which means in Afrikaans 'foresight'. The small farmhouse still exists and still looks

*Vooruitzigt was the original spelling.

out on a beautiful view. It turned out to be particularly well-named for it was owned by the two De Beer brothers, and it was their name which was going to be adopted by the greatest diamond mining company in the world.

The De Beer brothers were tough, close, bearded farmers and they beat off all approaches by the smooth-talking financial gentlemen. For a year they let out the diggings themselves, and all the time agents for companies and spokesmen for syndicates hung around their farm and followed them about and knocked on their rough front door. Then in 1871, pestered to compliance, they finally sold out for £6,300 to a syndicate of financiers up from Port Elizabeth. It was not a bad profit they reckoned, for the farm had initially cost them only £40.

If their two neighbours had been pleased by their sales the De Beer brothers were trebly delighted. Someone asked them what they would have done with the extra money they must have gained by waiting even longer. 'Bought a new span of oxen', replied one De Beer swiftly.

But they had sold out for peanuts. For a mere total of £10,000 three farming neighbours had parted with the world's richest diamond fields. Twenty years later £10 million could not have bought them.

The diggers now spread over all three farms and the tented cities grew and merged together. The sun sizzled and the packed veld danced with choking dust. Then icy rain lanced down like bayonets. Mud sucked at cold feet. Tents leaked. Men shivered and shook.

An American mining engineer Gardner Williams from Michigan, who was to become General Manager of De Beers, arrived on the diggings and set down what he saw:

'There was the oddest medley of dress and equipment: shirts of woollen – blue, brown, grey and red – and of linen and cotton – white, coloured, checked, and striped; trim jackets, cord riding breeches and laced leggings, and "hand-me-downs" from the cheapest ready-made clothing shops; the yellow oilskins and rubber boots of the sailor; the coarse, brown corduroy and canvas suits, and long-legged stiff, leather boots of the miner; the ragged, greasy hats, tattered trousers or loin cloths of the native tribesmen, jaunty cloth caps, broad-brimmed felt, battered straw, garish handkerchiefs twisted close to the roots of stiff black crowns, or tufts of bright feathers stuck in a wiry mat of curls; such a higgledy-piggledy as could only be massed in a rush from African coast towns and native kraals to a field of unknown requirements, in a land whose climate swung daily between a scorch and a chill, where men in the same hour were smothered in a dust and drenched in a torrent.'

CHAPTER ◆ SIX

The start of 'The Big Hole'

It soon became plain that there was another basic difference between the alluvial diggings along the river valleys and the rich 'dry' strikes of the 'pipes' on the farmland: the latter were obviously going to last much longer. And so, as the diggers prospered, tents gave way to buildings.

To start with there was no brick. Corrugated-iron roofs and corrugated-iron walls on wooden frames with holes for windows made the crude shacks. They were drier than the tents, but only a little cooler in the heat, and slightly less freezing in the winter nights, 4,000 feet above sea level.

Dealers set up offices, not in blocks of concrete and chrome but in one-room tin shacks. In came tradesmen who built stores to keep the goods they had hauled from the distant sea. Up went what their owners with a grand deception called 'hotels'. In fact they were crammed doss-houses creeping with bedbugs and rats, where bunks were double-used day and night and many stood out in open backyards.

As the tin shacks simply replaced the tents as they stood, and as the tents had sprung up in disordered confusion to be near a man's claim or his mates, no plan existed for the new town. As they had been in medieval Europe, roads were simply linked spaces winding between single-storey buildings and drained on both edges with wide and dirty ditches. Across these stinking dykes planks gave access to doors under lean-to roofs of corrugated-iron.

Kimberley today follows the same irregular plan of streets shooting off at angles and round bends, a layout to which all Europeans are accustomed, but which surprises South Africans used to neat towns plotted out in rectangular blocks, using plenty of space.

As the town grew up it needed naming. The inhabitants called it 'Kimberley' after the then Secretary of State for the Colonies, the first Earl of Kimberley, or John Wodehouse as he had been born in 1826. He had been Lord Lieutenant of Ireland and had been rewarded with his peerage in 1866 for his good work there in 'John Bull's Other Island'. The diamond diggers had cause to be particularly grateful to him. At an uncertain time when control over the whole area was in dispute, he had acted with bold foresight. It was by his direction that Britain gave her protection in 1871 to the mines on the tattered fringes of the new town.

British ownership came about in a curious and, her enemies complained, in a typically perfidious manner. The discovery of great riches and the arrival of 50,000 new inhabitants in a former backwoods area had aroused sharp interest from the two Boer Republics and one English colony bordering the place. First, President Pretorius of the Transvaal Republic, eldest son of the great Boer leader Andries Pretorius, made his claim. The land, he declared, belonged to the Transvaal, and he intended to come down and inspect it.

His grandfather (in whose honour the Transvaal's capital Pretoria was built and named) had inflicted in December 1839 a terrible revenge on Dingaan, King of the Zulus, at the battle of Blood River. 3,500 Zulus without firearms but hurling batteries of spears were slain vainly charging round and round a square of Boer oxwaggons. These, bivouacking by the river's edge, were defended by 700 picked marksmen sworn to revenge their leader Piet Retief, the great Voortrekker who, with fifty friends, had been butchered in cold blood when guests in Dingaan's regal kraal. Blood River extinguished the power of that extraordinary Zulu nation which had held sway over vast areas and several million other Africans, and yet had never invented either the wheel or an alphabet.

President Pretorius boldly announced that the digging areas north and west of the Vaal belonged to his Republic. Furthermore, he granted to a Boer Company a twenty-year exclusive lease over the diggings, subject to a 6% royalty to his Government.

The enraged diggers at Klipdrift formed the Diggers' Mutual Protection Society and elected as their President a tough British ex-sailor, Theodore Stafford Parker. Such a determined front did the new Society put up, and so menacing were its members when President Pretorius visited the diggings with an armed guard, that the Boers had prudently to withdraw when the diggers threatened to lynch the self-styled Chairman of the new Boer Company.

Less belligerently, because they stood more in awe of Britain and Cape Colony, and possessed a more subtle leader, the Government of the other Boer Republic now stepped forward. President Brand of the neighbouring new Orange Free State (recently Britain's Orange River Colony) now laid claim to the area occupied by old Nicolaas Waterboer and his Griquas, which unsurprisingly included the diamond workings.

By now Cape Colony had persuaded the British Government in London who were only distantly interested in South African affairs, to give their tacit approval to an expansion of British interests northwards over the diamond fields. They gave as their reason – and their cynical enemies cursed their hypocrisy – that 'any Boer extension in the area would probably lead to much oppression of the natives and disturbance of the peace'.

But the Cape's cause was not unjustified. From the beginning of the South Africa story the Boers, as compared with the British, generally treated the Africans and Coloureds under their control with Calvinist strictness.

London agreed to appoint an arbitrator of the boundary dispute, but the man selected could hardly be said to be unbiased: Sir Richard Keate was Governor of Natal, the other British Colony. He was, London declared, the only administrator of top rank in southern Africa not directly involved in the argument. After deliberating over the claims, Keate decided that the Kimberley area belonged to the Griquas. A new state of Griqualand West was declared. The Griquas, who preferred the British to the Boers thanks to their earlier support against the Orange Free State, now openly threw in their lot with the British. In 1871 Griqualand was declared a British Protectorate and part of the Empire. The Union Jack flew over Kimberley. London had got what it wanted without fighting.

President Brand found this operation hard to stomach and kept sending in plaintive bills for the Free State's compensation until his claims were finally paid off in 1876 by a settlement of £90,000. If honour was not completely satisfied at least the small agricultural republic had received a reasonable payment, and had postponed further armed conflict with the British.

But Waterboer, the unfortunate Griqua leader, received a painful setback simultaneously with the Free State award. The Judicial Enquiry decided that Waterboer himself had no claims to the diamond fields whatsoever. The British could be said to have come out best from their handling of the boundary dispute.

In the fields, the diggers had to evolve a new method of recovery. No longer were their claims spread out along miles of a river's banks, affording easy access. Claims were now concentrated on 'pipes' with diameters of only about 500 feet. The claims were at first limited to squares of 31 feet, and these therefore crossed the 'pipes' all over as tightly as Tudor panes in a lattice window. When digging began, however, the level panes became like the cells of a honeycomb.

Each little gang dug away at its share of the yellow ground in the pipe while those on the perimeter were finding nothing. Then, as the dozens of separate pits bit deeper and deeper into the centre problems flared. Because gangs dug at different rates one central pit might be 20 feet deeper than its competing neighbours. How did you get the earth out? And how did you prevent your neighbour's walls falling in on your plot below?

Temptations also loomed: you saw a diamond's pale gleam in the flank of your neighbour's plot on your very fringe. If you'd dug a bit straighter it might have been in your claim. Your neighbours were busy digging away up top ... Accusations of theft, of claim-jumping, flashed across the diggings and it was remarkable that, before the existence of any police force or any government control, violence and murder were not as widespread as in the American West. Perhaps British influence insured that fisticuffs settled most problems without escalating to knives and guns. Certainly a man in early Kimberley who could not use his fists had little chance of surviving, and Barney Barnato, the poor East End Jew, would probably not have become a multi-millionaire in the Diamond Fields if he had not been a dab hand at the Noble Art.

Just before he and his future rival, the Empire-builder Cecil Rhodes, arrived on the Fields, a fresh diamond strike on the edge of Kimberley aggravated all problems. 'De Beers New Rush', as it was first named, started only 2 miles from the first strike on the De Beers brothers' land. This time, instead of being in a 'pan' or slight depression, the pipe burst out of the top of a 'kopje' (which is pronounced 'koppie' and means a small hill in Afrikaans). Because the find was made accidentally by the native servant of a group of young men prospecting from Colesberg, the field was first called 'Colesberg Kopje'.

The party was led by the splendidly named Fleetwood Rawstorne, who had an old Bantu manservant called Damon who was usually drunk of an evening. One night Rawstorne found him particularly inebriated and kicked him out of camp to dig away on the hill to work off the liquor. The party were camping on the kopje as a base for prospecting round about. No one imagined they were sitting on diamonds. But there in the moonlight Damon spotted a gleam and snatched up a stone and raced back to his master who was finishing dinner with his friends round the camp fire.

Unfortunately it was Sunday night and a keen Sabbatarianism kept one day free from toil and gain: no claims were even allowed to be registered with the Diggers' Committee on the Sabbath. Yet if they waited, word would certainly leak out and they might get beaten to the claim. 'The Red Cap Company' as the Colesberg party called themselves after the red woollen hats they all wore that crisp winter, whispered the news only to a few close Colesberg friends. The word might as well have been publicly circulated by that new device, the electric telegraph. They had hardly prepared their oxen at daybreak for the short haul to the find when the veld came alive around them and they were pursued on to the site by a rocking armada of straining oxwaggons, goaded on by greedy rivals.

Because it was on the edge of Kimberley the new mine came to be called after the town. Its relict, the famous 'Big Hole', the greatest pit dug by man without sophisticated machinery, still looms black, dark-green, grey and horrid in the centre of the modern city.

You see from the viewing platform the bands of rock cut as if out of a model: the red top soil, the basalt, black shale, melaphyne and quartzite as the great hole dives down into the earth to its bottom lake of turgid, pea-soup, sinister water. Of the yellow ground and the blue ground nothing, of course, remains; only the great shaft looms where the diamondiferous pipe struck downwards.

This time the diggers had learned from previous difficulties. The Diggers' Committee, with the democratic zeal engendered by communal effort before personal rivalries emerge, had been given considerable powers by its members. It now laid down that 15-foot roadways had to be left right across the area every 47 feet. This meant a severe loss of valuable ground, but it afforded easier, safer working and access for everyone – at any rate to start with.

The yellow ground was dug out of each claim into little buckets which were hauled onto the access tracks either by ropes or by workers climbing up steps cut in the pit sides. At the top waited each man's waggon, mule or pony cart. When these were filled with earth they pulled out for the separate washing and sorting bases which ringed the perimeter of the workings like supply troops around a fighting army. There, with dearly bought water, the soil was sluiced through the cradles and then painstakingly sifted through by hand on the sorting tables.

Sharp-eyed supervision was thus needed in three places: on the claim, *en route* for washing and during the sorting. In the first years more than half the diamonds prised from the ground were being stolen by the native labourers. While all the diggers were engaged full-time on their own claims there was no possibility of overall policing. Deterrents were instigated, but detection was difficult unless you gave up time from digging. And security control, the basic safeguard by which workers are isolated for the whole term of their labour, could not be implemented while a host of individual diggers employed a mob of different workers.

But there was now a new problem. While fierce regulations, many of which are still in force, were being drawn up against Illicit Diamond Buying, the weight of the carts and waggons on the narrow tracks was causing landslides. Earth, equipment, oxen, horses, mules and men tumbled over the slippery sides on top of the diggers sweating down below. The pudgy yellow diamondiferous ground made a shifting underpinning for roads. The situation in the deeper diggings became impossible; soon the more central claims were almost cut off.

The solution hit upon was to haul from the perimeter. Wooden platforms were erected like scaffolds on the edge of 'The Great Hole'. From these, long ropes stretched out and down towards each separate working. At the bottom was a bucket; at the top a winch. Up squeaked the rope over the scaffold's cross-bars and hundreds of feet away the laden bucket started its journey to the crater's lips. To reduce the tangled confusion, the platforms, called 'stagings', stood in three ranks on ground level. Those closest to the edge hauled buckets from the diggings directly below them; their ropes hung almost vertically downwards. The second rank served the central diggings and the rear rank, standing highest of all against the sky, hauled the earth right across the workings from the far side.

Everybody in the end had, perforce, to use this method and so the great, deepening crater was woven and warped with hundreds of ropes making a giant spider's-web over the digging ants. At the surface the stagings bristled black against the horizon, silhouetted like siege-machines of an old investing army.

Winching was strenuous work and as speed was of the essence, the richer diggers dispensed with human haulage and powered their hoists with horse and mule. Round great wooden wheels called 'whims' the animals padded, and the turning wheels shortened the ropes and brought the buckets jerking up to the surface far more quickly than human hoisting.

Speed, however, still called. Each day's living at Kimberley was hideously expensive. The diggers could not work out their claim too quickly and those who had recently sold some good stones invested in steam engines to get their buckets really whirling up to the top. They had to hope for a good haul of diamonds daily for coal in Kimberley cost more than £40 a ton, a year's wage for a servant then.

As the shafts sank deeper, damp oozed in underfoot. Then the digging tapped the underground water table and the rains drummed down on what had now become a gaping catchment area. The claims started to flood. The base of the Kimberley mine became first a quagmire, then a lake. Frantic hand-pumping could no longer bale the water out more quickly than it was rising. The end of the mine as a workable proposition loomed in sight. Already some diggers despaired and moved off to prospect elsewhere.

But there had arrived in Kimberley in 1872 a 19-year-old Englishman, son of a vicar, descendant of English farming yeomen, who had been forced to come out to Africa for his health's sake. This youth observed the flooding, thought of a steam-driven pump, considered where he could find such a thing on the South African earth, finally traced one in distant Port Elizabeth, negotiated its purchase on loans, had it hauled up to Kimberley and hired it out to the diggers. It was in constant demand. The diggers paid through the nose to use it. Its purchase price was so swiftly repaid that it was possible and profitable to buy more.

The foundations not only of a fortune but of an empire were laid this quickly when young Cecil Rhodes arrived in Kimberley for a shot at the diamonds, and had the vision to think of a steam-driven pump. The need was there, but so were thousands of experienced miners. It was the Hertfordshire vicar's sickly son who hit upon the cure.

CHAPTER ◆ SEVEN

The dreams of Rhodes and Barnato

The man who was going to make a vast new land out of the wilds of Africa was born in the vicarage at Bishop's Stortford, Hertfordshire on 5th July 1853. It was a plain white-fronted late Georgian house, large enough for the clergyman's extensive family, but in no way grand, though there were, of course, domestics and there was some entertaining. Many vicars of the Church of England were at that time the younger sons of aristocrats; their elder brothers inherited the estate or went into the Army. The Reverend Mr. Rhodes was not of that class, but he mixed with them.

'The greatest of living men', his son Cecil was to be called by Rudyard Kipling. He left in the wake of his brief, fierce life a host of fervent admirers and many bitter foes. Both factions would agree that he was a visionary, that he thought on a world scale: a railway from the Cape to Cairo, the Union Jack over half of Africa – 'All that English; that's my dream.' He was called 'The High Priest of Optimism', yet he was so ill with the consumption which finally killed him, that he was thought unlikely to live through his twenties.

He grew up a pale, delicate boy, with wide-apart eyes and, until its shape was later disguised by a huge moustache, a large, sensitive and melancholy mouth. Little is related about his mother other than that she was the second wife of the Reverend Mr. Rhodes and that she bore him clever children. This son was brought up by a Nanny who found him a handful.

Cecil was the youngest of five brothers; all the others became soldiers. It seemed at first probable that he would follow his father into Holy Orders. He would have been an unusual parson but could conceivably have made a brilliant Pope. He was to rule a large Empire as a benevolent autocrat, brooking no criticism and never doubting his infallibility.

He was a highly competitive schoolboy at the local Grammar School; 'a disinclination to lie behind any of his rivals whether in their studies or their games' – thus was his character described by a contemporary biographer, Howard Hensman. Hot-headedness, impatience of authority and intolerance of argument all stamped his childhood and hinted at the man. Once, seizing a heavy book, he was within inches of using it to assault a young master who, he decided, had punished him unfairly.

Nobody recalls spotting in him the early flickerings of the flame of genius, though his old nurse remembered that he was wilful, adventurous, disobedient and brave: he climbed out of windows on prohibited night expeditions which were dangerous to his weak chest. Rhodes never did accept his physical limitations.

He had very little to do with women and appeared deliberately to shun their company. The outside world knew nothing of any woman in his life and the attempts of London Society hostesses to lionize him when he achieved fame were all rejected. Some of his associates declared that he feared and hated women. His friends on the other hand believed that he refused to fall in love lest the loss of emotional control might play havoc with his career. Some of his descendants, however, deny he was a misogynist, citing his whispered relationship with a German baroness, allegedly his mistress, to whom he gave a valuable string of black pearls

He was a ragbag of contrasts: a huge man, yet when roused by impatience or zeal for a plan, his voice piped out in shrill eunuch squeaks. Sick in his lungs, he grew up with the outward appearance of health; a pink complexion, clear eyes and rich wavy hair. He was a stern administrator but his eyes were often aloof and reflective. Burning with energy, he yet walked slowly and contemplatively, hands often back on his hips, fingers fanned downwards. But his big nose and strong chin displayed his aggression and tenacity, and he leapt at decisions like a hound at a fox.

His plan was to leave Bishop's Stortford Grammar School for Oxford, but at sixteen his health broke down so severely that his local doctor feared he would die. To go to Oriel College, Oxford, was a potent childhood's dream, whetted by his frustrating illness. He held to it always, accomplished what he wanted – the life there, not so much the degrees – and remembered Oriel with loving munificence on his death.

If he was not to die as a child his physical salvation (and his financial fortune) lay faraway in a warm dry climate. Cecil's eldest brother, Herbert, was farming cotton in southern Natal, which though not dry, was certainly very warm, and on 21st June 1870 he set out on the long voyage in a small sailing vessel to join him. Durban, Natal's only port, had been the scene of brisk skirmishing, sieges and reliefs first between the early English settlers and the Zulus and then between English and Boer. By 1870, however, it was firmly British.

Rhodes was thus, if not exactly on the spot for the diamond rush, at least within hundreds of miles of Kimberley and on the same land and not 7,000 sea miles distant in Britain. His sharp ears caught the fresh chattering enthusiasm about the new discoveries. Back in Britain this talk was thinned by time, diluted by distance, and largely discounted. News from the diggings came racing east into Natal and many of its settlers cast aside farming in that colony for a crack at the diamonds in the new land beyond the Drakensberg and the Orange Free State to the west. Elder brother Herbert soon found the lure irresistible, and off he went. A few months later his reports were so exciting that young Cecil could wait no longer. He had some money, his health had been greatly restored in Natal, and he had grown impatient with cotton farming.

He had no notion of either spending a long time in Africa or of making a fortune. His prime objective continued to be that of attending Oriel College, Oxford. For that he needed good health and some money. He now had the former. Perhaps the new diamond fields would produce the latter.

In 1872 Rhodes arrived at Kimberley to join his brother. There he fell to, a young prospector working a joint claim, digging himself, supervising their native workers, living rough, sorting diamonds and selling them to dealers. He and Herbert shared a single claim at first and worked hard at it for good rewards but without sensational success. He wrote home to his mother about finding a stone of 17½ carats, so that must have been an outstanding stone on their claim. He told his mother he was paid £500 for it – 'an absurdly high price'. Rhodes decided to widen his activities. He developed ancillary occupations selling water and ice from an imported ice-making machine to the diggers.

The drier air of the high interior plateau was even better for him than the climate of Natal. His health bloomed. The claim was producing a steady income before he launched his lucrative pumping business. At 19 he was fit and rich. But so far from feeling poised at the start of a new career in a new land, he resolved to go home to ancient Oxford.

And that might easily have concluded Rhodes' African adventure if his health had withstood England's cold, clammy hand, and if he had not, first, made a ruminative journey on his own.

He did not go straight back to England. Instinct told him to take a last, long look. He spent eight months in a solitary journey through unmapped country to the north of the Orange and Vaal Rivers. He travelled in or beside an oxwaggon walking 20 slow miles a day across the empty, rolling veld. He completed a vast loop through Bechuanaland to Pretoria, on to Middelburg and then back across the Transvaal to Kimberley.

He looked and he pondered on this lonely peregrination. He saw the genesis of a fine rich healthy country. He guessed the Kimberley diamond fields to be only the first signs of many other rich mineral fields. He formed before he was 21 his colossal life-plan: to bring first Africa, and then the world under British influence. In his first Will written when he was 22 he set out this aim and when he died he left his fortune to further it.

Rhodes went to Oxford and matriculated, but then in 1873 he collapsed again. 'Six months to live', wrote his doctor. On a last hope Rhodes sailed again for southern Africa and journeyed up to Kimberley.

Great changes had been made. Fifty thousand inhabitants were living in a solid town and the £100 claims of the early days were changing hands for £4,000. Observing the competition, it seemed to Cecil Rhodes a dreadfully wasteful way of mining. He decided that the amalgamation and consolidation of claims was the only efficient way of operating them. Through his pumping operations he was already connected with the De Beers mine and that was therefore the base on which he resolved to build. He was 20 years old. Back home in Britain his fellow undergraduates were thinking of philosophy and tennis parties, poetry and politics, girlfriends and strawberry ices.

He had formed his pumping company with two other young Englishmen, Charles Rudd and Wallace Alderson. The latter was an engineer by profession, and Rhodes found himself in the role of financier in the partnership. Though all three partners had put money into the business he found he had to buy his small claims in the De Beers mine – as he'd bought his first steam pump – by borrowing on forward Bills at a few months' notice. He wrote home that while Alderson was dreaming about engines his own nights were plagued with scraps of paper with his signature on them looming up too soon for payment.

The original rules limiting claims to only one per digger soon collapsed under economic pressure. The limit was extended up to ten claims per person. Consolidation was beginning at the bottom as some diggers left for new fields and others collapsed and sold their claims to their neighbours. With the money flowing in from his pumping business Rhodes took over claims all around him, sometimes by direct purchase, sometimes by first going in as a partner. Though it was going to take 7 more years before he became one of the biggest claim-owners in the De Beers mine he was already a young man of consequence in Kimberley when his direct opposite and future rival, Barney Barnato, arrived on the scene in 1873.

It was a bad autumn for the diggers who had given little thought to the future marketing of their treasures. It was hard, living rough out in the wilds, to think of distant European dealers and international price-structures, when quick profits lay in your grimy palm and pressing bills had to be met each day. When diggers found a stone they usually sold it that day for the best price offered by any dealer, passing merchant or 'kopje walloper'. With their sale, cash in sovereigns and notes, they bought themselves food and paid their rent and their claims' rights. They paid their natives and mended their roofs and bought champagne, a mule, a woman, a cartwheel or a new pick and shovel. Via Kimberley and Cape Town the diamonds were shipped off to Europe and America from whence all the money seemed wonderfully to flow for ever.

Then in the autumn of 1873, financial crises broke out like thunderstorms first in Vienna and Berlin and moved like typhoons through Paris, New York and London. Rich banks crashed and the market for diamonds halted in a week. In Paris and Amsterdam nobody bought a diamond. In Africa the diggers toiled away and no one bought a stone. Bankruptcy loured across the diggings.

Onto this grim scene bounced Barney Barnato. Barnet Isaacs – to use his original name – was the son of Isaac Isaacs from grim

Whitechapel, grandson of a Rabbi, educated at the Jews' Free School, part-time variety artist, a bit of a boxer and cocksure he'd strike it rich. No novelist could have created a better contrast to match against the mighty Rhodes.

Barney and his elder brother Harry boxed together under father Isaac's keen tuition: the two lads must learn how to take care of themselves in a tough world. The Jews in Britain were still socially ostracized at most levels. The Prince of Wales, whose friendship for and indebtedness to many leading Jewish families established them among the aristocracy, was not yet a power in the land. Neither Isaacs brother was keen on the evening studying which might have got them clerks' posts in the City. Barney adored theatres and would hang about their doors begging part-used tickets off people leaving early at the interval. At first he used the seats himself. Then he started dealing at cut rates in half-used tickets. He was a speculator early.

Next he wanted to act, but could find no theatre to take him seriously. One West End manager offered him 6d a day and a hot meal to turn cartwheels. Barney's first whiff of greasepaint wasn't exactly Hamlet: he wore a heavy moustache, acted the clown, and Harry joined him in a double act, playing the gentleman to Barney's buffoon. 'Barney too!' Harry used to shout to get him onstage. When the brothers wanted a stage name with an Italian ring they hit upon a corruption of the cry: 'Barnato'. Under this new billing Barnet Isaacs changed his name for ever.

West End theatres, however, still did not enthuse over the Barnato Brothers. Their earnings were precarious. Barney's disappointment plunged into discontent. Then their cousin David Harris looked in one evening in 1871 to see them. He was babbling about the news of the diamond rush on the Vaal River in distant South Africa. The idea gripped Harris; he worked on his mother, extracted almost all her savings of £150 and set out for Africa. Shortly afterwards cousin Harry decided to follow.

Young Barney, at 20, was left seething behind. His parents had not the means to finance him, but he had saved nearly £100 himself from his ticket-dealing and acting, and a century ago third-class steerage to Cape Town from either London or Southampton cost only fifteen guineas for the 5,951-mile journey, advertised as 'The Shortest, Quickest and Cheapest Route to the Diamond Fields'. When Barney

Barnato embarked in July 1873 on the brand-new 2,000-ton steamer *Anglian* it broke the previous record to the Cape by making the journey in 28 days. To boost his bounding confidence he took with him a nickel watch presented by his school-friends, and 40 boxes of dubious cigars given to him in partnership by his uncle Joel Joel. They were only half a present for they had arranged to split the profits on their sale.

Barnato met his first bad news as soon as he docked in Cape Town. Rumours of difficulties in the diamond fields had travelled south, spurred on by those who had failed up there, and by farmers who, having sold out at the top, could now knock the market in idleness. Barnato was too late. His mercurial temperament slumped. He had blundered, he was sure. But his pride prevented him returning to London defeated, so he took the train as far as it went, which was to Wellington a mere 45 miles inland to the north-east.

There he missed a 6-horse coach and a 12-mule waggon. The first went without him and the second asked £40 for the journey, more than twice as much as the ship for only a tenth of the distance. It was an impossible price for him. Gloomily he hung about for nearly a week. Then he settled for a waggon with four oxen, the humblest transport going. This charged him £5 to carry his baggage, so long as he walked most of the 580 miles at its side.

It was early September and the African winter was ending in pouring rain. Barnato slept the cold nights sodden and exhausted beneath the dripping axles. By day his party trekked dourly onwards across the Karoo, searching for the winding track and straining to push the wheels through mud. To the normal basic African fare of 'mealies' (maize as porridge or paste) was added the Boer's travelling meal: *biltong*, strips of dried salty meat which taste to an Englishman like the soles of ancient shoes.

The oxen plodded northwards at 3 miles an hour questing across country to find better going and the shallower 'drifts' or fords. Forty miles crept past on a good day, and spring began, but when they reached the Orange River it was flooded. The baggage and stores were loaded into native boats and ferried across, so that the lightened waggon could make a shallower crossing upstream. But nearly a week had been lost.

Now they began to be overtaken by richer transport trotting forward to the diamond fields. They also met the returning poor pedestrians who had failed. Barnato's unstable spirits soared and dived with each new traveller on the now busy track.

After nearly two months' journeying they spotted the outlying jumble of tents of Dutoitspan crowded and steaming on the horizon. The waggon creaked forward into the mud street of wooden and iron huts. Barney began to enquire about his brother, searching gloomily through the shacks till he found Harry lying hunched up and hungry in a squalid hut.

Harry's claim had shown rapidly diminishing yields. Cousin David Harris was doing equally poorly and now, to cap all, the international economic crises in distant Europe and America had stopped all sales. The situation appeared desperate.

Barney showed a brave face, took his brother out to an eating house called Martin's Hotel for dinner, and winced when the simple meal cost him £2. More than 6% of his working capital squandered in an evening! But Barnato was resolute. If the cursed diamonds had now failed, he must try something else. First he tried Payne's Circus by standing up to challenge the professional champion, a moustached Portuguese from Angola. Barnato wore his smartest London clothes (he fancied very narrow trousers), a bowler hat and had hastily doffed his glasses. The crowd thought he was a riot, the Portuguese treated him as a fool, but Barnato had not forgotten his early skill. He knocked down the big man from Angola and the rough crowd cheered. But there was no long-term money to be earned there. Even when he added jokes and songs from the ring the crowds no longer paid to see 'Barney Barnato, World Famous Comedian' and for even 5s a day the circus could not afford to keep him. Barney lasted four days and was paid off by Mr. Payne.

Then he tried peddling, buying the cheapest rubbish from caravans freshly arrived from Cape Town and taking notebooks and pencils round the claims. Without the money to buy even part of a claim Barney's only way into business was by dealing. He would try to be a 'kopje walloper', a fellow riding round the claims buying stones from diggers on the spot, then taking them into the townships for a quick sale at a small profit.

To learn about the look and feel of diamonds Barney asked some diggers to let him give their ground a second washing and inspection. Men with good diamondiferous claims yielding decent stones could not waste time rechecking their earth for small fry so they let young

Barney do it for a few shillings – 'findings keepings'. To give him a hand he employed a native whom he had noticed enthusiastically scribbling nonsense with one of his unsold pencils. So he paid the man in pencils to haul up the water buckets to him. He found almost nothing: a few tiny stones at most. But he made many contacts and some friends, he learned methods, he valued stones, he watched the 'kopje wallopers' haggling and he carefully noted prices.

Barney had come across another very young Jew named Louis Cohen, who already had some experience of 'kopje walloping'. He was not, however, as Barney observed, a good salesman, so Barnato convinced him of his need to combine with a 'thoroughly expert dealer' like himself. Barnato's enthusiasm was infectious. He proposed a partnership to Cohen with the money each possessed. This combined would give them a modest 'bank' for diamond dealing. Cohen put in £60 but Barnato got away with £30 plus the 40 boxes of dubious cigars his uncle had given him in London. The two new partners found a solicitor to draw up the agreement between '*Loo Cohen born in Liverpool aged* 20, *and Barnett Isaacs known as Barney Barnato born in London aged* 21.'

Next, decided Barnato, they must open an office. But a mere hut in Kimberley's main street cost £1,000. Barnato approached the publican who owned Maloney's Bar to rent him a lean-to at its side. The Bar stood between the Dutoitspan and the Kimberley diggings and Barney had noticed that it was the evening's first port of call for many diggers from both areas.

They rented the hut for a guinea a day. This meant their combined capital, if spent on nothing else, would give them a tenure of a mere 90 days. They had staked all on a very short haul and not an hour must be wasted. They moved their sleeping bags, a table, chair, weighing scales and a lens into the miserable 9 foot by 6 foot shanty. The future millionaire Barnato was in the diamond business.

CHAPTER ◆ EIGHT

Clash of giants - Rhodes versus Barnato

While Barney stumped round the claims buying stones, Cohen kept the office to sell them. Then Barnato brought off a master-stroke. Exhausted by his long days' trudging and frustratedly aware of his restricted radius he kept wishing he could afford to buy a pony. Then he heard of one owned by a 'kopje walloper' who roamed the distant claims where the diggers were out of touch with the high prices obtaining at the centre. Because few buyers troubled to ride out to see them, the outlying diggers sold their stones more easily for less.

But the best was to come: Barnato learned that the old pony could pick his own way round the outlying claims and find his way from one distant digger to another. That pony he must have! He fell into talk with its owner, convinced him he wanted the pony for someone else who would not be a competitor, and, after hours of wrangling, obtained the pony, saddle and bridle for £27 10s 0d, nearly half of the little partnership's remaining capital. Yet after a few days travelling with the astute pony the profits on Barney's resold diamonds had recouped his pilot's cost. Now he had the supplies, profits came easily. He was a skilled dealer.

Meanwhile Loo Cohen was blundering away in the shanty office. Barnato was a climbing man and poor Cohen was a failure. The ambitious are usually ruthless: Barnato dissolved the partnership and instead, and according to normal Jewish practice, joined brother Harry and cousin David Harris. The new partnership, using their old stage name of Barnato Brothers, was registered at the end of 1874, and the unfortunate Loo Cohen vanished like jetsam in Barney Barnato's wake.

As Barney travelled the area he was sniffing out opportunities. Against the consensus of local opinion, he was convinced that the deeper blue ground would hold diamonds exceeding in both quantity and quality those more easily shifted from the yellow ground above it. The yellowish earth normally ran out about 60 feet below the surface and when diggers struck the hard slate-blue formation beneath it, they assumed that profitable work was over. The honest and foolish abandoned their claims; the dishonest and equally foolish covered the bottoms of their pits with yellow earth and sold out to ingenuous new prospectors.

Barnato took advice from the experts and particularly from the Government geologist at Grahamstown, the same Dr. Atherstone who had identified the first 'O'Reilly' diamond found by van Niekerk.

Barnato decided that, as diamonds were created by a combination of heat and pressure, the deeper the 'pipe' from which they sprang the larger the stones should be. He was not necessarily right over the size of the deeper diamonds, but he was abundantly right about their quantity. He was equally right about concentrating on the centre of each mine.

There were four adjoining holdings in the centre of the Kimberley mine belonging to the Kerr brothers. Because the Kerrs believed their area exhausted they offered their four holdings at £3,000 compared with the £40,000 the combined claims would have made if only half exploited. £3,000, however, represented Barney and Harry Barnato's total accumulated capital and Harry required prolonged encouragement from his ebullient younger brother before he was persuaded to embark on the colossal gamble.

Almost as soon as their natives' picks dug into the deep blue ground diamonds were found. Up they were hauled to the surface for washing and sorting, each day bigger, each day more of them. The stones were soon coming up at the rate of £3,000 worth a week, and Barnato was coming up to be one of Kimberley's richest men.

Like Cecil Rhodes over at De Beers mine, Barney knew that claims must be consolidated for efficient working. Rhodes' claims in De Beers bordered those of Charles Rudd, the Old Harrovian who was Rhodes' partner in their pumping enterprise. Their shared middle-class and educated background had initially brought them together; now they combined their holdings and were to remain lifelong friends. Years afterwards Rudd related that he, Rhodes and a third partner Alexander Graham had been offered the whole of the De Beers mine for £6,000 in those early days, but had not been able to afford to buy. On 1st April 1880 however, they merged with three large rivals to form the De Beers Company with a capital of £200,000.

When Rhodes first formed 'De Beers', Barnato had already bought the valuable Kimberley claims of the Standard Company. Now Barnato absorbed the Kimberley Central Mining Company and expanded his empire again. But by the following year, 1881, landslides in the Kimberley mine had grown so terrifyingly frequent that few men dared work under the constant threat of death under débris. No sooner had Barnato unearthed the treasures in the deep blue ground than his competitors aped him all around, blasting and burrowing greedily into the earth without respect for life.

When the Barnato Brothers Diamond Mining Company was formed in 1880 with a capital of £115,000 a minority of the shares had been offered to the public. Since the £3000 patch had already yielded diamonds worth £250,000 the public had stampeded for shares and their price rocketed. Now, as digging and recovery became more difficult and dangerous, they began to fall again.

In the De Beers mine landslides were not to become a problem until 1886, so that while Barnato was embroiled in fighting collapsing shafts in his Kimberley mine, Rhodes at De Beers was pushing quietly ahead with further expansion.

The Barnatos now took on more of the family. Their three Joel nephews, Woolf, Jack and Solly, recently arrived from England all joined the enterprise. Then Barnato, compelled to avoid the landslips from outside, evolved a system of driving shafts into the ground right outside the mine's perimeter and then cutting into the 'pipe' by long underground passages at different levels. These workings set the basic pattern for the deep mining of diamonds today and were accomplished without modern machinery in a few years; by 1884 Kimberley deep mining was in full production.

Thus far the paths of Rhodes and Barnato had yet to cross. Both men knew and respected each other and saw in each other the only rival for control of a gigantic mining combine. For this was Rhodes' immediate target, a combination of the four mines; De Beers, where he now held full control, to be added to Bultfontein and Dutoitspan – and to Kimberley where Barnato was winning sole control.

Like any imperialist nation on the road to expansion, the force of Rhodes finally hit a rival power. War was inevitable.

Of the two contrasting protagonists, Rhodes, the tall, stooping, aloof Oxford man with the pink face, was far more ambitious than the little dark bespectacled Jew from London's East End who had left school at 14 and developed a mania for showing off. Barnato suggested a division of interests to Rhodes: that he be left alone to control all Kimberley, while Rhodes would remain the unchallenged master of De Beers. So do world powers dispose of spheres of influence.

But Rhodes wanted a monopoly, not a duopoly, for an additional reason to that of efficient *production*. Cecil Rhodes wanted efficient *marketing*. He saw that the demand for diamonds would grow and grow as girls grew up and got engaged and had a diamond ring to prove it. He envisaged the spread of wealth downwards through the democracies, so that the grand-daughters of house-maids who worked a whole year for £100 would wear diamonds on their engagement fingers. His rival's nephew Solly Joel saw it the same way and said pithily, 'While women are born, diamonds will be worn.'

But it was particularly necessary to control the marketing of diamonds because of the uncertainty of the precise number of engagements likely to occur at any time. The supply of diamond rings must not be too little so that sales opportunities were lost, nor so large that the price fell. Monopoly alone could control the balance. To create a monopoly battle began between the two giants.

Rhodes attacked first with an attempt to buy a parcel of claims in Kimberley. He was rebuffed: they went to London rivals.

But a French group owned a large belt of claims splitting the holdings of Barnato's Central Mining Company. When Rhodes heard that Barnato was about to bid for Compagnie Française de Mines de Diamond du Cap de Bon Espérance, he was desperate to prevent him. Rhodes had a German friend of his own age, Alfred Beit, who had arrived in Kimberley in 1875 from Hamburg to represent a diamond-buying firm. Beit's partner was another German, Julius Wernher who had recently fought in the Franco-Prussian War. Together they formed the firm Wernher, Beit and Co.,* which held

*The firm was to become world-famous. Julius was created a baronet in 1905. His nephew Sir Harold Wernher, G.C.V.O., the 3rd Baronet, married Lady Zia, daughter of a Russian Grand Duke and a kinswoman of the Queen. The Wernhers live in considerable style in their enormous country house, Luton Hoo. Alfred Beit's descendant, Sir Alfred, the second baronet, lives in one of Ireland's largest and most beautiful country houses, Russborough. Both families may be said to have risen a long way since the rough days of Kimberley only 90 years ago.

stakes in all four mines including a holding in the French Company.

Cecil Rhodes and Barney Barnato were each worth well over £100,000 when they joined battle, but Rhodes had the advantage of having Alfred Beit as an ally. The Hamburger greatly admired the tall Englishman, and shared his vision of a monopoly. Rhodes reciprocated his respect. Indeed his admiration for Germans in general continued all his life. (His attitude may surprise us today; it was normal thinking in Rhodes' time. Prussia, Hanover and other German states had been Britain's allies for two centuries. It was France then who was Britain's historical foe.) So Rhodes was convinced that the Germans and the English constituted the world's two finest races, and he gave explicit expression to this belief in his Will which set up the Rhodes Scholarships at Oxford.

Beit's strength lay in his many valuable contacts in the world of banking: one was the powerful Lord Rothschild. Ironically it was to this aristocratic Jew that the Hertfordshire vicar's son turned for finance in his struggle with the little Whitechapel Jew. Rothschild and Rhodes got along famously together. Sir Nathaniel Rothschild, Bart., was made a British peer in 1885, but had been born the heir to a Baron of the Austrian Empire. The family's financial empire covered Britain, France and Germany and their immense fortunes had been given a multi-million pound fillip in 1815 when, by dexterous use of carrier pigeons and special messengers from the field of Waterloo, the Rothschilds learned of the Anglo-Prussian victory well ahead of their financial rivals.*

Rhodes first wrote to Nathaniel Rothschild setting out in a memorandum his sweeping plans for the new diamond industry of southern Africa. He received no immediate reply. The Rothschilds' policy was of non-involvement in areas where there seemed a probability of strife. They distrusted the Anglo-Boer quarrels. Their instinct told them that serious trouble lay in wait. They felt their money would be too far away and too much at risk in Africa. Their adviser on the spot, however, a man named de Crano, was a friend of

*Nathan Rothschild himself met the packet-boat at Folkestone and beat Wellington's own envoy to the Government by several hours. He went straight to the Stock Exchange and proceeded to *sell* Government stock. His rivals imagined he had learned of England's defeat. Panic selling sent the stock crashing. Then Rothschild bought a gigantic line of stock for a song. He had made a fortune before the dead were buried at Waterloo and while victors and vanquished lay dying together.

the American mining engineer Gardner F. Williams, who was an associate of Cecil Rhodes. The wheels began to mesh. Rhodes played this link hard, for Williams had a good financial brain as well as being a top-class engineer. Rhodes set him to work on a full report on De Beers and this was forwarded to Rothschilds via de Crano.

Rhodes followed the report with a personal visit. He sailed for England with Gardner Williams to call on Rothschilds. What he was seeking was a loan of no less than £1,000,000 to buy out the French company. Deploying Alfred Beit to Paris there to ask the French company's board exactly what they would take for their holding, Rhodes in London opened simultaneous discussions with Nathaniel Rothschild. His charm, his passionate belief in his plans for a monopoly, his confidence in a great future of economic expansion, and particularly his faith that co-operation between Boer and English would come about and lead finally to concord, won over the great Jewish banker. A united and rich South Africa ... Like all great men Rhodes could see beyond his life-time. The unification of a rich and powerful South Africa lay beyond the bitterness of the Boer War and the slump of the early 1930's. But Rhodes' long-term vision was correct.

With the million-pound loan agreed, Rhodes hurried to Paris to meet the directors of the French company. They agreed to sell out for £1,400,000, and were prepared to take £700,000 of the sale price not in cash, but in De Beers shares. This suited Rhodes excellently: £300,000 of the Rothschild loan would then be available for additional working capital. The Frenchmen had to call an extraordinary general meeting of their shareholders to confirm the sale, but that, they assured their guests, would be a pure formality. Rhodes, Beit and Williams counted the great deal well done and returned to South Africa. There was, however, a dangerous delay before ratification.

When Barnato heard of Rhodes' proposed deal, he knew it would finish his control. There, under his nose, on his own pitch this confounded empire-building rival was seizing a slab of rights which would split his holdings apart for ever. Barnato struck back desperately. He was a shareholder in the French company himself, and he had friends who also owned shares in it. He persuaded them to unite with him to beat off Rhodes' offer. But the Frenchmen would have to gain significantly if they were to switch at the last moment to a new bidder. No petty increase in price would suffice. If Barnato put in a really big bid it could woo the French and silence Rhodes. He tempted the French company with a colossal rise: he offered £350,000 more than Rhodes' bid.

The magnitude of this offer was such that even those shareholders in the French company who had been strong supporters of De Beers and Rhodes, now switched their allegiance. Kimberley crackled with speculation. The two giants were fully stretched and one must crack. What could Rhodes do to recover his position?

Rhodes called on Barnato. His plan was first to threaten, then to seem to weaken and then to be prepared to negotiate. He told Barnato that with his backing in London he could afford to top his bid with £300,000 and another £300,000 and another *ad infinitum*. 'We shall have the French company in the end.' Thus the threat.

Then in a more reasonable tone he asked Barnato why their bidding should be allowed simply to benefit the French shareholders. Barnato wanted the French company. Very well. Let Rhodes buy it at his first agreed figure of £1,400,000 and then he, Rhodes, would sell it on to Barnato. Barney sniffed about feverishly for the catch. Rhodes added quickly that Barnato could pay him partly in Kimberley Central shares. They discussed the valuation and finally agreed on the double deal. Rhodes agreed to sell his newly-bought French company on to Barnato for £300,000 in cash plus one-fifth of the shares of Kimberley Central.

It was one of those good business deals when both sides retire happy. Barnato had his precious French company and was therefore, he thought, out of danger. But Rhodes was a move ahead. He now had what he really wanted: a powerful foothold in Barnato's Kimberley Central. This, expanded and exploited, could lead to his dream: total control over the diamond industry so that production could always be equated with demand.

What had seemed peace was only a lull before the last ferocious campaign. Rhodes wrote afterwards: 'We had to choose between the ruin of the diamond industry and the control of the Kimberley mine.' He had to, as he said, 'get the whip hand' over 'obstinate people'.

His attack was two-handed, like the gladiator with trident and net. With one hand, and backed by Beit, he prodded Barnato forward by buying Kimberley Central shares to add to his holding. With the other he dangled the net: he accelerated diamond production from his De Beers mine until it equalled, then surpassed the production from Kimberley.

The harassed Barnato was thus doubly menaced. He saw his precious majority in Kimberley being whittled down daily as more and more people – including his treacherous friends – sold out to Rhodes. To preserve his control, he had to enter the bidding himself, buying as Rhodes bought, chasing the shares which should have been allied to him anyway. The price of the shares soared higher and higher. Rhodes' trident had prodded Barnato far enough. The dangling net was ready for the final offensive. Rhodes started to flood the market with his diamonds. He undercut Barnato lower and lower. The price of diamonds sank below production cost. The price of Kimberley shares was thus precariously far too high. An immense crash threatened.

Rhodes called again on Barnato and let a bagful of diamonds roll across the table. 'That's only one day's production,' he said. He showed Barnato that he had £2,000,000 worth of financial backing from Rothschilds, Wernher, Beit and other bankers. Finally early in 1888 Barnato could stand no more. He surrendered. He would sell out to Rhodes, and take payment in De Beers shares. De Beers at last owned Kimberley, and the new company was formed: 'De Beers Consolidated Mines, Ltd.' So it remains to this day. 'Head Office: Kimberley. London Office: 40 Holborn Viaduct, E.C.1.' Rhodes had won his great monopoly.

One final scene concluded the drama. Some minority Kimberley shareholders objected to the deal and took the case to court, saying their company's Articles of Association only allowed it to be merged with another mining company. De Beers was more, they claimed. It was a huge financial empire. 'They can do everything,' their counsel pleaded. 'Since the East India Company no company has had such power as this.' The judge agreed. The shareholders won their case. The sale was illegal. This meant no more than a mere delay to Rhodes and Barnato. If the Kimberley company couldn't be sold as an entity, they would liquidate it and De Beers could then immediately buy its assets.

So it was done that way and the legal trouble deftly ducked. And that is why the framed cheque in De Beers' boardroom in Kimberley today is made out to 'The liquidators' in the sum of £5,338,650. At the time it was the largest sum in the world's history ever written by a company on a cheque. The date was 10th July 1889. Rhodes had been seventeen years in Africa.

CHAPTER ◆ NINE

"A Fancy for Making an Empire"

In the sixteen years since young Barney Barnato had walked exhausted into the shanty town called Kimberley, the place had grown into a boom city. A municipality was officially constituted on 27th June 1877, the year in which Griqualand West was taken over by Cape Province. In 1880, when Rhodes formed the De Beers Company, it had its third mayor, Sir Joseph Robinson. It opened a waterworks, and two years later installed electricity. Kimberley's staggering streets were 'Illuminated by Electric Arc Lamps', and by 1885 the railway line had been driven up from the Orange River to reach the new town. The settlements ringing the Dutoitspan and Bultfontein mines had grown into a town too. Later to be merged with Kimberley, this was constituted in 1884 as 'Beaconsfield' after the title taken by the British Prime Minister Disraeli from a Buckinghamshire village.

These stirrings of local government showing the influence politics could usefully bear on commerce had not been lost on Cecil Rhodes and Barney Barnato. Both entered the political arena via the Cape Parliament. But Rhodes was first an undergraduate at Oriel College, Oxford, keeping the terms of which he had long dreamed, and handling the build-up of his financial empire during the Long Vacation between June and October. Few other 23-year-olds can have built such foundations between terms. Rhodes did not in fact study hard at Oxford. It was the life, and perhaps the position which he had always craved, not the pursuit of learning to gain a degree. The quality of his companions changed rapidly between Oxford and Kimberley. In term-time he hunted with the university Draghounds along with all the young bucks from Christ Church and the squires from the 'shires'. He read a lot – mainly the classics and particularly Aristotle – but was rebuked by his tutor for not going to lectures. He passed his exams in the end, but by the time he returned for his final term to take his degree after another long break in South Africa he had become a member of the Cape Government. A rare type of undergraduate, he won the seat of Barkly West (formerly the old Klipdrift river diggings in 1881 at the age of 28 and he was to become a Prime Minister at 37.

At Oxford he went on spinning dreams of Africa. He dreamed particularly about a great expansion to the north through the Bechuana territory which abutted the diamond fields. In 1881 it was still a no-man's-land through which passed the trade routes up The Selous Road through the magically named Khama's Country, Matabele Land, and Mashona Land up to the Zambesi River. This huge heart of Africa was then ringed by German and Portuguese territories on the east and west, and across the north by the Congo Free State.

The southern artery to this dark heart of Africa lay through Kimberley. Pressing against the artery was the Boer Republic of the Transvaal. The 1881 Pretoria Convention had tried to stem any further westward expansion by the Transvaal, and had stopped encroachment just to the east of the Selous Road trade route. Rhodes served on the boundary commission and persuaded Chief Mankoroane, who claimed half of Bechuanaland as his country, to cede his land to the British Cape Government. Rhodes' efforts were wasted. The Cape Government, reluctant to take on any further commitments and preferring the comfort of rest to the risk of movement, flatly rejected the projected cession. The British Empire, built on trade, was hardly ever helped by British colonial governments. What would have been a rewarding step northwards under British protection stumbled at first stride.

In 1884, while his rivalry with Barnato was waxing hot in Kimberley, Rhodes contrived to find the time to act as a Resident Commissioner in Bechuanaland. The Boers there had broken the Conventions, driven out the Africans and set up two new pocket republics of Goshen and Stellaland. President Kruger of the Transvaal denied, as governments tend to when turning a blind eye helps their interests, that he could control these freebooting commandos. Their illegal activities were helping the Boer cause, as Queen Elizabeth I's privateers had helped England against Spain in the sixteenth century. In South Africa it took the dispatch of the Warren Expedition – a British force poised and menacing on the Transvaal frontier – to convince Kruger that he could withdraw his roving commandos.

The Cape Government, fearing the possibility of another war, now supported Rhodes' plan for expansion to block the Transvaal's forays. The Union Jack would, after all, march northwards. It was a good example of the start of many of the Empire's local expansions. The possibility of trade brought with it the likelihood of trouble. Where bright-eyed local British traders spied out the former, reluctant British governments had to act to avoid the latter. 'Trade follows the Flag' was seldom true: the Flag usually followed to support the traders. So Southern Bechuanaland was declared British territory after all and its northern area became a British Protectorate right up to the 22nd parallel.

British access to the immediate north was assured, as Rhodes wanted. But it was only the first step, and not far enough. The fires which were going to blaze into the South African War between Boers and British had been held back, not damped down. Next year, 1886, gold was discovered in the Transvaal's Witwatersrand, and the flames leapt up again, fanned by conflicting greed.

Cecil Rhodes' political efforts had the same impetus as his business ones. He sought expansion under one flag, which was, surprisingly, not the Union Jack. The flag he envisaged was a South African one representing a federation of united Boers and British, governing themselves within the framework of the British Empire. He died just before his dream was accomplished, but the unlikely Union was going to last for nearly half a century, until a close referendum after two World Wars converted the Commonwealth's Union of South Africa into an independent Republic.

After De Beers Consolidated signed their £5,338,650 cheque to the Liquidators of Kimberley Central, Rhodes' path lay towards the Premiership of Cape Colony, and he was regarded in Britain more as a politician than a magnate. His new admirers wished he need not concern himself with commerce. His new business partner, Barney Barnato, disagreed fiercely: he wanted Rhodes to keep to business. Their new company was valued at about £23½ million. It controlled 90% of South Africa's diamond output, and its first year's trading yielded a net profit of £1,000,000. Financially the colossus was the fantastic success for which Barnato had prayed. But as Rhodes saw it, a diamond empire was another step in power politics. When Britain finally realized the wealth and potential of the diamond fields, she would grow seriously interested in South Africa. Rhodes' plans for a united nation and expansion northwards would at last secure British backing.

Seven years after Rhodes became Member for Barkly West, Barney Barnato decided to follow him. In 1888 when Barnato lost control of his Kimberley Central to Cecil Rhodes' De Beers, Rhodes persuaded him to run for the Cape Parliament, as Member for Kimberley. Barnato, however, treated politics much as an extension of his passion for acting. He generously supported Kimberley's little local theatre so that he could play all the main roles, even subsidizing a charity performance for the local hospital to the tune of £10,000 to enable his small frame to play the giant Moor in *Othello*. When a professional actor sulking in the front row dared to snigger at his performance, Barnato leapt over the footlights in his robes, dived into the stalls, punched the mocker over the head and then clambered back onstage to continue his speech. The crowd loved it.

For politics he hired a coach-and-four and a team of outriders and postilions, all in matching livery of green and gold and scarlet. He dressed up his personal aide-de-camp in a scarlet frock coat and top-boots so that he looked like a circus ringmaster. For himself in a dusty new town in the middle of Africa he chose the dress for Ascot's Royal Enclosure in a sophisticated English June: grey morning coat and top hat. But it lacked panache: he added light blue lapels. Electioneering was equally music hall. Whisky was freely stood all round, Barney released a spate of funny stories, the diggers thought the whole performance a splendid joke and returned him cheerfully to the Cape Parliament.

CHAPTER ◆ TEN

Smugglers, Whores and Clubs

After the consolidation of the mines the seeping sore of Illicit Diamond Buying could be properly dealt with. In the early days theft was rife on the diggings: half the stones dug up each day were smuggled out and sold. By the time of the De Beers merger, the institution of security compounds, a special department of police and detectives empowered to search rigorously and a disciplinary tribunal capable of meting out savage punishment, had reduced the rate of loss. But one stone in twenty was still getting out to be sold illegally on the Black Market.

The IDB chain usually started with a rich entrepreneur safe in the background, who was often based as far away as Cape Town. He would be a man of considerable means, outwardly respectable for those times and parts. (Much passed in the rough, tough frontier colony which would not have done in Park Lane.) The background entrepreneur financed the whole project from the first theft of the stones right through to their final export to Europe.

He was serviced by the original receiver – usually a storekeeper on the diamond fields to whom an African 'runner' brought the stolen stones. The 'runner', sometimes employed in the mine or on the workings, was the link with the Africans actually in the mine: by 1891 there were 6,000 African workers underground. The stones did not always come out the same day they were sighted. When the 'blue ground' was lying out weathering and awaiting sorting, a native would mark a diamond, hide it there, and go back days later. The stone was first flicked up between the African's toes so that nobody could spot him stooping. Then it might go into his hair or ear before it was swallowed. Diamonds were also hidden in gaps between teeth or behind glass eyes. On the next stage between the dealer and master mind they were smuggled inside hollow books and in the false heels of shoes.

The quickest and safest way out from the mine to the dealer was inside an African's belly, and sometimes there were four or five sizeable stones (the dealers could be choosy: they didn't want small stuff) rattling around inside a thieving worker.

The external scrutiny of Africans leaving the mines was already so ferocious in 1891 that the *Pall Mall Gazette*'s correspondent remarked how degrading it was: 'It's lucky you have black labour handy,' he commented. 'No white would stand this sort of thing for any wages under the sun.' Further indignities awaited the suspected swallower of stones. His hands were first handcuffed into fingerless leather gloves and then he was purged with castor oil and stewed dried fruit and confined in a room under strict supervision until the dose had, after several griping hours, taken its full and foul effect.

The great deterrent to IDB was the crisp enforcement of the Law which is still in existence and which absolutely bans the possession of any diamond unless bought through a licensed dealer. Possession of a diamond in Kimberley today bears the same grave responsibility, and visitors to the city are seriously warned of this. The onus of proof is on the possessor: he must prove he came by a diamond legally.

Under a unified South Africa, prevention, arrest, search and discovery is reasonably easy. But eighty years ago 'runners' could slip quickly from the diamond fields in Griqualand West into both the Transvaal and the Orange Free State. The latter had strict anti-IDB regulations, but no machinery at all for enforcing them. In the Transvaal Republic and the English colony of Natal few questions were asked, but there still remained the long sea route from Durban and Cape Town to Southampton. Getting the illicit diamonds home on their last lap to London required a final subterfuge. There were a number of attractive and generous English ladies who earned uninhibited and generally horizontal livings on the diamond fields. Often they seemed to need to sail home pregnant. Their corsets were stuffed with diamonds for the voyage back.

The unification of the mines under Rhodes and Barnato meant an end to casual labour. From then on, all African workers employed were kept in security compounds. The same system, much refined but even more strictly controlled, obtains today. A *Pall Mall Gazette* correspondent gave an eye-witness account of workers in the Kimberley compound in 1891.

'A "compound" is a spacious quadrangle built round with iron sheds, a sort of combination barracks and exercise yard, in which the Kaffir employees of a company are kept during such time as they spend above ground. Kaffirs, under supervision, perform all the unskilled labour of the mines; and under the "compound system" they contract with the company for three months at a time to work and board and lodge. During that term they are practically prisoners in the compound. There half their time is spent, sleeping, eating or playing games. The other half is spent in the mines, where they gaily sweat away underground from six to six, on bread and soup and "mealies" for wages at the scale of £1 to £2 a week. To and from the mines, between the pit mouth and the compound, they are marched under vigilant escort. Within the compound they are cut off, for the term of their service, from the outer world and from all drink stronger than ginger beer ... All the other small needs of the Kaffir, food and clothing as well as drink, are supplied at cost price within the compound walls. From quarter to quarter he is "taken in and done for" in sickness or in health. Disease and drink are the sworn enemies of dividends.

'And for disease and drink the raw Kaffir, as he first totters on to the diamond fields, is fair game. He has tramped, belike, hundreds of miles from the far-off huts of his tribe – from Basutoland, or Zululand, or Gazaland – hundreds of miles with no food but what he carried. Weaklings have sunk and died upon the way; but he is strong, and he has crawled on to the goal. His tongue hangs out for lack of water, and his belt is pulled tight for lack of food, and he staggers into Kimberley on such a pair of skeleton legs as it will take months to fatten in the compound even with double rations and light work.

'But to see the same man nine, six or only three months afterwards, retracing the same road with his face set homeward! Sleek and well set up, he stalks along with the air of a returning hero. He is the proud master of a rifle, a blanket and a red umbrella. He has also a leather belt well lined, and that means cattle, and that again means wives. He has made his fortune out of the white man, and he goes home to enjoy it.

'It was a Sunday when I visited the De Beers compound, and the natives were amusing themselves at their own sweet will. It accommodates some 2,000 of them. Some were lying about in the queerest mixture of European costumes; their Sunday togs. One lounged in a hat and a waistcoat, not a rag else; another strutted in an old pair of uniform trousers with a red stripe, a very dirty shirt hanging over them at large. Others sunned themselves in native nakedness. Eager groups were intent on a kind of faro, sovereigns passing freely over the cards, for they are great gamblers. A Shangaan put on his feathers and danced a *pas seul* for my express benefit. A band of Zulus – one splendid young savage among them I well remember – stamped a wild war-dance till the sweat poured down their dusky frames, and each man had stamped himself some inches into the ground.'

The compounds are much smaller now, but they still resemble the better sort of barracks and offer good food and clothes at subsidized prices. They now present piped music, football, athletics and free cinema shows. There is less nakedness and fewer war dances. There is still an unsated demand to get in, workers still travel hundreds of miles to queue for jobs, and they still return physically fitter and better fed after their contract ends.

The obliging Kimberley ladies of the 1890's who smuggled the diamonds out belonged to the cohorts of prostitutes who buzzed about the growing city like bees making honey. In the earliest crude days so few women were available that, like a big Cockney barmaid nicknamed *La Singularité*, they auctioned themselves publicly every evening, assured of a good price for the night's work when demand so boldly outstripped supply. One night *La Singularité* sold herself for 25 sovereigns plus three cases of champagne. The wine had cost at least £70 by the time it reached Kimberley so the barmaid cost the winning digger about £400 in terms of modern money.

Supplies soon improved. Brothels sprang up to cater for hungry clients and there were soon nearly a hundred in the town.

Crime decreased as stronger forces of police and detectives moved into the area. The diggers still decided some issues with their fists and allowed claims to go to the stronger of two, regardless of merit, but firearms were still hardly used.

Some sporting instinct had been brought from England where 'gentlemen still played the game'. Carrying a pistol, let alone using one, was considered 'bad form' in Kimberley from the earliest days. There were surprisingly few murders or armed robberies of the type which were enlivening the American West. Sleazy parts of the town existed like London's Soho, but in the main businesses, stores and new hotels opened up, and the banks moved in with the doctors, solicitors and accountants to fill the needs of a respectable urban community. Sports grounds opened: the first Currie Cup rugby match was played in Kimberley in 1892, when a World Exhibition was held in the public gardens. A racecourse flourished, promoted initially by Barney Barnato. Dances took place. A society was born. And the Kimberley Club was started.

Britons, when they possessed their Empire and settled round the earth, always set about forming clubs. A need to belong and to conform, a need for male company and a need to exclude inferiors were strong English traits which have still not died among the Clubs of St. James', on the rugby field, or in the uniformed gangs on motor-bikes.

The 'Kimberley Club affair' exactly posed the differences between Cecil Rhodes, a basic clubman of sound background, and Barney Barnato, the sort of noisy extrovert whom good clubmen loathe and yet who always craves to be elected. In Kimberley Rhodes was one of the Club's founders. The other founders were only small-fry gentry too by London's standards, but they were Kimberley's new society: they had evolved certain standards; they all had some position; they liked each other; they did not like the bumptious little Barnato, who was not, they thought even by colonial standards, a gentleman.

The Club, officially formed on 1st August 1881, sprang from a group of Kimberley's leading citizens who, calling themselves 'The Craven Club', used to meet, dine, drink and chat in the Old Craven Hotel. The members then acquired a clubhouse of their own. The Club's premises on the Dutoitspan Road have been twice burned down. The present building is one of considerable dignity, rich atmosphere and some unexpected contrasts. The airy balconies fronting the street on both floors do not prepare the visitor for the marvellously Victorian hall bristling with antlers, glowing with dark furniture and faintly lit by a huge stained-glass window halfway up the grand baronial staircase. A sort of lost Britain has been preserved in waterglass on the unlikely veld.

The Club has always been business's social heart in Kimberley. Everyone executively engaged in the city's industrial life seeks membership and most obtain it. And because diamond mining is *the* industry and De Beers *the* company, the latter have often carried the burden to keep the Club running.

Among the legends attaching to Cecil Rhodes and Barney Barnato on the diamond fields is the suggestion that Barney's election to the Kimberley Club was his *sine qua non* of consent to merging Kimberley Central with De Beers.

There is no evidence of any secret understanding between Africa's two richest men of that time, but it may well have been implicit. Barnato needed urgently to join the Kimberley Club; Rhodes was its official founder and prime mover. If anyone could get Barnato in it must be Rhodes. The two former rivals had come to terms. Rhodes, so often austere, went to exceptional trouble to please his new colleague. He knew that public acclaim was Barnato's breath of life. Thus he suggested the new title of 'Life-Governor of De Beers' which so delighted Barney.

Rhodes had encouraged Barnato to follow him into the Cape Parliament and had had to support him afterwards. Barnato's membership was not universally popular: the poorer diggers (of whom there were still thousands) suspected that what Barnato represented was not so much Kimberley as De Beers. Privately, Rhodes and Barnato acknowledged some truth in this, but they could have argued fairly that Kimberley really *was* De Beers. From the diamond fields around it was now coming half the annual earnings of the whole of Cape Colony.

Rhodes' new kindnesses to Barnato were not entirely altruistic. His dreams of a British Empire stretching into Africa's heart had flourished and were now becoming plans in which the wealth and power of De Beers had a vital part. Diamonds efficiently mined and marketed would give him the finance and influence for his next long stride. But Barnato regarded the business of diamonds as an end in itself; political preocupations were entanglements. Rhodes would need all his fabled charm to woo him round.

Steps had already been taken. The discovery of gold on the distant Rand in 1886 had confirmed Rhodes' prognostication of Africa's huge and hidden mineral wealth. The following year his associates C. D. Rudd, 'Matabele' Thompson and Rochefort Maguire held a farewell lunch with him in the Kimberley Club before inspanning their oxen behind the Central Hotel and setting off on the 700-mile trail up to Matabeleland. There they obtained concessions over all the Matabeles' mineral rights from King Lobengula in the royal kraal at Bulawayo. Rhodes was ready for the next step.

The great expansion needed De Beers' support and this required Barnato's blessing. Rhodes took Barney frequently to lunch in the Kimberley Club. Barney was so proud of these invitations that he made sure all the town knew, as he peacocked along Dutoitspan Road on his way to the Club. But a club rule forbade a visitor to lunch more than once a month. And so, as the gossips said, Rhodes finally 'made a gentleman' of Barnato and got him elected.

Barnato in his turn supported Rhodes' project. In June 1890, Cecil Rhodes became Prime Minister of the Cape, and the first Pioneer Column set out from the Club for Matabeleland. King Lobengula had unwittingly signed his nation's death-warrant. He agreed to a partial occupation of his land by Europeans engaged in prospecting and by their accompanying police escorts. Out of these private police grew the semi-official force of the British South Africa Company which in its turn became Rhodesia's army. For this expedition the famous guide Selous led a party of 700 settlers and police from the British South Africa Company. The first seeds of Rhodesia were about to be sown.

CHAPTER ◆ ELEVEN

The Gold Crash and the Jameson Raid

There was further progress in the diamond fields. The Diamond Syndicate was established in 1893 and the next year the railway, financed by De Beers, was driven up from Sterkstroom to Indwe.

Eastwards, high on the Rand in the Boer Republic of the Transvaal, the interests of the two great men in the diamond industry now expanded into gold mining and its ancillary enterprises. Rhodes and Barnato were about to become politically and financially involved in a double disaster which respectively ruined one reputation and amputated one life. South Africa's future richest city, Johannesburg, was growing among the goldmines. Growing too were the old squabbles between the resident farming Boers and the powerful, numerous and enterprising group of Uitlanders of all nationalities who now lived in the Transvaal. Positions were being reluctantly taken up for the future war against a background of soaring values. During 1894 the value of Rand shares quoted on the London Stock Exchange trebled to nearly £60 million. Stockbrokers across the world from New York to St. Petersburg responded to the feverish cries of their speculating clients: "Buy more 'Kaffirs,'" as South African gold shares were first nicknamed. Little was known in distant Europe and America about the new mining and investment companies mushrooming in the Transvaal. Barney Barnato, the diamond millionaire, seemed to be behind the best of them. Stockbrokers abroad followed his lead.

Then, on the Jewish day of Yom Kippur in 1895, large blocks of 'Kaffirs' were suddenly sold on the Paris Bourse. In the absence of all Jews stock markets everywhere were unnaturally quiet, and there were few buyers around in Paris for the avalanche of shares on offer. In an hour the shares lost a third of their value. Now anxiety was telegraphed to other drowsy stock markets across the world. Presumably some hideous disaster must suddenly have hit the Rand. Panic selling started. Prices plunged. Small investors lost every penny they had put in, and many rich men who had wildly bought shares on borrowed money were now trapped, desperately selling shares to raise funds as the prices crashed to the bottom. Many accusing fingers were pointed at Barney Barnato. He, the prince of diamonds and king of gold in South Africa, was in England running his racehorses and building his great new house in Park Lane. He must be responsible for the crash. Weren't the two stone statues outside his new mansion called 'The Petrified Shareholders'? Hadn't this pint-sized charlatan recently walked on his hands down the dinner-table at a stuffy white-tie-and-tails function at the Savoy to remind his shocked neighbour that he'd once been a clown? The Establishment in London muttered anxiously to one another: how could you trust a feller like that?

The market slide continued. But Barnato's volatile temperament bore up surprisingly calmly. Not only did he disregard public criticism but he refused to join the panicky rush of selling. He then took steps to bolster up the faith of the financiers. He went about loudly expressing his supreme confidence in the Kaffir market now and for the future. The City of London was so grateful for his support that it arranged a banquet at the Mansion House for the former Barnet Isaacs of Whitechapel, where his confidence could be publicly reiterated and he could be thanked by the Lord Mayor.

His noisy optimism could however only slightly ease the financial depression. Political strife was growling loudly in Johannesburg. There, nearly all the money, brains and enterprise came from the 'Uitlanders' – those foreigners, mostly British, who had moved in to develop the goldfields and quickened the expansion of all the supporting industries. But they remained foreigners in the Transvaal Republic. They had no electoral rights at all, and no municipal voice in the city which they had created and for which they were heavily rated and taxed. They were in truth supporting the Boer farmers whose land it was. The rancour of the 'Uitlanders' festered. Then Lionel Phillips, President of the Johannesburg Chamber of Mines, made a speech which shook the confidence of foreign investors.

'It is a mistake to imagine that this much-maligned community, which consists, anyhow, of a majority of men born of freemen, will consent indefinitely to remain subordinate to the minority in this country.'

Lionel Phillips was not only in a position to speak for all industrial Johannesburg. He was an associate of Alfred Beit and he belonged to Rhodes' political faction, so his words were particularly significant. In Europe and America gloomy views were taken. More bloodshed seemed likely. And in South Africa the mood was now one of dangerous action amongst the British. A certain Dr. Jameson and his supporters were about to end the great Cecil Rhodes' political career.

Dr. Jameson, a black-moustached, thickly-built gentleman of tremendous courage was Administrator of Rhodes' British South Africa Company, which had obtained from the British Government in October, 1889, an extraordinarily powerful Charter. In effect the Chartered Company administered and policed vast territories of Africa including Mashonaland (occupied in 1891 to prevent a Boer trek), and Matabeleland (conquered in 1893 to quell Matabele raids on their old foes the Mashonas).

From the northern border of Bechuanaland the Company's mandate extended through Khama's Country, north across Lake Ngami and the Zambesi River, past Victoria Falls and on almost as far again to the shores of Lake Tanganyika. Eastwards its territory bit deep into what was to become Portuguese Angola. Westwards it spread to Blantyre and Lake Nyasa and Portuguese land again. As Rhodes announced to his shareholders: 'We have a country 1,200 miles in length and 500 in breadth, and it is mineralized from end to end.' After six years Rhodes' efforts had subdued, unified and pacified several countries, and in May 1895 all the lands administered by the Chartered Company were officially and in his honour re-named 'Rhodesia'. Dr. Jameson, as the Company's Administrator was thus a man of singular power and a close associate of Rhodes.

With the most secret support of Cecil Rhodes, who was now Prime Minister of the Cape, several hundred prominent British businessmen in Johannesburg had been plotting for some time to overthrow the Boer Government of the Transvaal. Rhodes' motives for supporting an insurrection remained the same; he wanted to remove an obstacle to northward expansion. He therefore not only countenanced but encouraged the plans of the rebels. Arms and ammunition had been smuggled in to them and were hidden in the gold mines. They now planned the next ploy of an oppressed majority, as they, with justification, saw themselves. There were now 80,000 'Uitlanders' in the Johannesburg area, four times as many as the Boers. They would appeal for help to an outside force which would march in to liberate them as they rose in revolt. It was a classic usurpation ploy.

Their liberators would be armed forces of the British South Africa Company under Dr. Jameson. These were already poised on the Transvaal's border for they were stationed in the former British colony of Bechuanaland, recently incorporated by Rhodes into his own Cape Colony. Dr. Jameson, who had been accustomed to subdue 6,000 fierce Matabele or Mashona warriors at a time, envisaged little difficulty in brushing aside a few untrained farmers and burghers.

It was the first of many serious British underestimates of the courage and skill of their neighbouring Afrikaner settlers. The Boers had no option but to be resolute and courageous: unlike the British, French, Dutch and German settlers they had no other homeland now to which to return. Their ancestors had come to Africa a century and more ago. They had no ties now with Europe. They were a race built up of refugees from oppression. They had, in the Biblical sense which remains so strong in them, found their Promised Land. They were not to be evicted. The Uitlanders were monied people with business brains who dug up gold and diamonds and built fortunes. But the Boers were people of the country, knowing, because they loved it, how to move and fight in it and how to defend it from invaders.

The Raid, reckless in concept and incompetently plotted, was a fiasco. Jameson marched into the Transvaal with only 700 men, half the force he had intended.

Cecil Rhodes was gnawing the bone of anxiety in his lovely house 'Groote Schuur' on Cape Town's outskirts. He had sent his brother, Colonel Frank Rhodes, a lively cavalry officer, to take charge of the conspiracy on the Rand. The signal for the invasion or 'liberation' was to be an attack on the arsenal at Pretoria.

But delays were caused by ludicrous debates as to which flag should be raised by the Reform Committee once they had publicly revolted, and by reasonable concern about the official support the rebels might expect from Cape Colony or Britain. The first question was never decided and Rhodes kept on intimating in the last weeks before the invasion that there could be no promise of any official aid. Most of the pessimistic plotters would not continue without some such assurance, but the optimists argued that once the *coup d'état* was established, both Cape and British Governments would recognize it with public decorum and private joy.

In the event no attack on the arsenal ever took place. Jameson was ambushed at Krugersdorp 18 miles west of Johannesburg. The Raid was over. And to the south, a train load of the Uitlanders' wives and children en route for the safety of Natal was derailed with many dead and seriously injured. The cause collapsed.

Half the Johannesburg business men failed to act at all, and some scurried to Pretoria not to seize the arsenal but to apologize to the Transvaal authorities. The ringleaders were quickly captured. There were immediate ominous repercussions in Europe, for the German Kaiser telegraphed his congratulations to the Transvaal's President Kruger in words which plainly warned the British Government to keep out. The British Colonial Secretary had perforce to condemn the Raid.

Rhodes could no longer remain Prime Minister of Cape Colony. He promptly resigned. His dream of a united South Africa expired like a pricked Christmas balloon in a few January days. His career as a statesman was ended.

The little Transvaal Republic rejoiced. Not only had its farmers easily subdued a rising, but world opinion was up in arms behind it. It generously returned Dr. Jameson and his followers to the British authorities, but sent 63 of the Johannesburg plotters on trial for treason. A judge from the adjacent Orange Free State presided and condemned four of the ringleaders including Colonel Frank Rhodes, Lionel Phillips, and the American engineer Hayes Hammond to death. All the other plotters including Barnato's favourite nephew, Solly Joel, were sentenced to two years' imprisonment.

The death sentences raised protests round the world and *The Times* correspondent in Johannesburg reported that even there Boer citizens were signing petitions for them to be commuted. Mrs. Kruger herself wrote letters expressing her sympathy for the wives and children of the four condemned men.

Barney Barnato now took upon himself the last great role of his varied life: that of an international mediator. He flung all his fierce energy into attempting to obtain reprieves for the rebels.

Barnato had protested so volubly during the trial, mocking the judge and the prosecution, that only his numerous noisy supporters in Johannesburg's streets had saved him from arrest for contempt of court. When the sentences were confirmed he called on President Kruger, 'Oom Paul' as he was affectionately known to the Boers. The President was friendlily disposed to Barnato, who had brought great wealth to his little Republic and yet had until now never interfered politically. He listened carefully. Barnato's plea to save the four condemned men was long, eloquent and impassioned. But the President was obdurate. His anti-Uitlander faction were howling for the men's blood, recalling a British execution of five Boers at Slaagter's Nek way back in 1812. He could not, he said, interfere with the course of justice.

When his fiery words failed, Barnato's riches were summoned into action. The heart had been appealed to vainly; now came the inescapable logic of the head. Barnato, worth even then £12 million was the most powerful man in all the Rand. He reminded Kruger coldly, 'I employ 20,000 Europeans – as many as all your Burghers combined, I employ 100,000 Kaffirs, the work force of the city. My businesses spend £50,000 here every week ...' He would, he threatened, close all his businesses, remove all his money, dismiss his huge labour force, if the four men died. To prove his intent he advertised all his properties for sale the very next day.

The ground crumbled beneath Kruger's feet. He foresaw the economic collapse of his Republic which had in a few years been elevated from the produce of a few thousand farms to a rich industrial state, by the enterprise and capital of Barnato and his associates. Kruger envisaged the threat of riot from the 100,000 unemployed Africans. He imagined the fury of the 20,000 frustrated and impoverished Uitlanders flung out of work. He contemplated the collapse of the city if all Barnato's businesses closed ...

'Wealth can't break laws,' Kruger used to say, but he now allowed it powerfully to bend them. The four death sentences were commuted to fines of £25,000 each and the prison sentences to fines of £2,000 each. Honour and lives were saved and a small state which urgently needed foreign investment acquired £200,000 for a popular act of mercy.

Barnato was a fêted hero among the British residents. In London a cartoon portrayed him as a music-hall conjuror hypnotizing the

Transvaal President, while Joseph Chamberlain, the Colonial Secretary, goggled in the audience and muttered: 'How on *earth* does he do it?'

Victory won, Barnato tried to soften Kruger's feelings of resentment. He knew that 'Oom Paul' in his heart still kept alive his early dreams of being a lion hunter, for it was the President's boast that he had killed his first lion before he was 12 and his first African when he was 14. So Barnato commissioned a fine pair of sculptured stone lions and had them delivered to the President's house. The gift was accepted, but Barnato was never completely forgiven for having bent justice in the Transvaal.

The ability to forgive and forget is not always present in the Afrikaner character. Forgiveness is an unusual gift in any race, but in older countries it is easier to see a setback as a small wave in the long sea of history. Forgiveness too thrives in the rich soil of a country which can afford charity more easily than in a new small country creating itself against natural and human opposition. To the small man each setback seems a defeat, and every slight an injury. In the character of South Africa today one sees repeated not only the strengths but also the weaknesses exemplified by old President Kruger, who was very much a man of his race.

As Barnato was to die so soon, it would have been nobler if he had died in the glory of his greatest victory. He had saved lives, the probability of civil riot, the possibility of war and even of a European conflagration. He had survived the loss of £3 million in the Kaffir market crash and was still twelve times a millionaire. But, stretched to the limit through his last two crises, his confidence like strained elastic began to fray and flag.

He was exhausted by his struggle with President Kruger and the Transvaalers. He looked with deep depression at the stirring troubles in Africa. To the north there were African risings and his old colleague Rhodes, disgraced after the Jameson Raid, was no longer on the spot to control the trouble. The British Government had confirmed his resignation.

Seven crammed years ago Barney had voiced his grave doubts about Rhodes' plans. After one of those long sessions in the quiet of the Kimberley Club where the maps of Africa fluttered in the light and Rhodes in confidence measured his phrases or squeaked out in excitement, Barnato had said to him:

'You have queer ideas. Some people have a fancy for this thing and some for that thing; but you have a fancy for making an empire! You want the means to go north, if possible, so I suppose we must give it to you.'

He had been reluctant to agree. He was sure now that he had been right to doubt, utterly wrong to consent. He had never liked politics. Back to business, he resolved wearily, and in the middle of 1896 he sailed back to London.

He was drinking heavily now. He had always been able to put away more than most men in those rough days on the diamond fields before Kimberley was built, only 21 years ago. But now back in London, wrestling with the complexities of his financial empire, he drank for the dangerous reasons: to create energy, to ignite enthusiasm, and to hire hope. He worried ceaselessly and usually quite unnecessarily. His companies had survived the great stock-market crash and the imminent threat of war, and were being brilliantly managed by his two Joel nephews, Woolf and Solly.

There was no real cause for concern, but black despair clutched him like a crow. He no longer started his thrumming day with a boxing bout. He no longer went regularly to his beloved theatre. Yet all his life he needed human response: this is what made his world smile. Now he shrank back as his world literally darkened. He was losing his sight. He groped miserably about the hated Mayfair house. He had never wanted it, never been happy in it. He loathed its mocking statues of 'The Petrified Shareholders'.

Outside it he gave money to every blind beggar as a premium to fate to insure his own failing sight, sometimes running back in the crowd to trace one blind man he had overtaken. Other nightmares possessed him: that every hand was plotting against him, that he could trust neither his staff nor his family. Occasionally the old fresh fire leapt up again in his fuddled brain and he worked furiously, but only in a fit. Gloom and fear returned as regularly as the dark night.

He sailed once more for South Africa, but only on a quixotic, last-minute impulse. He had intended seeing Solly Joel off from Southampton, then resolved to sail with him as far as Madeira there to take a holiday, and finally he decided to go all the way to the Cape. His spirits rose when he saw Table Mountain. New projects swirled over his brain like the cloudbanks wreathing the mountain-top. When he reached his headquarters in Johannesburg his mind was in an electric thunderstorm. He was gabbling so fast that he became unintelligible. His friends watched his disintegration with dismay. It was the height of Johannesburg, or perhaps its thin air? Or perhaps poor Barney was tormented by the Parliamentary Enquiry into Rhodes' part in the Jameson Raid? That corpse was not yet buried. Back in London the former Cape Prime Minister was facing the Committee of Enquiry in the House of Commons.

So Barnato left the high breathless plateau of the Rand and went down to the lovely Mediterranean climate of the Cape. The nightmares which had spurred him through Johannesburg's sleeping streets screaming, 'They're after me!' at friends' doors, now abated. For a while he was calmer in longer spells and even took his seat again in the Cape Parliament, making one speech which showed him quite at his best. Then tides of depression came roaring in across his mind again: the world had risen against him. His wife and nephew Solly tried to persuade him to return to England. But he saw persecutors lurking even in his close family: they were plotting to prise him out of his businesses ......

And then in 1897 came the grand excuse for a journey 'Home', as English-stock South Africans called Britain until recently. Queen Victoria's Golden Jubilee was going to be celebrated all through the Empire. Every patriot rich enough to travel should go home to wave a flag. A party of South Africans under the Cape's new Prime Minister, Sir Gordon Sprigg, were sailing for London in the SS *Scot* on 2nd June. Barney's patriotism brooked no delay. He would go home with them, accompanied by his wife and 3-year-old son and Solly Joel.

All went so well till they were 4 days away from London that none of the other passengers suspected his recent torments. His family, however, expecting a relapse, kept him under close watch. The 14th June on a calendar reminded him that it was the anniversary of his obtaining the release from prison of Solly and the other Johannesburg plotters. At luncheon his wit kept his table laughing and afterwards he enjoyed a lively talk with the Cape Prime Minister.

A little later, after strolling together round the deck, Solly Joel suggested to him that he sit by his side in a deckchair. 'What's the time?' Barnato asked. They were his last words.

As his nephew replied, 'I make it just 13 minutes past 3,' the wretched Barnato, presumably believing 'They' were after him again, dashed to the ship's side and started to struggle over the rails. He was going over as Solly reached him and grabbed Barney's coat. But the nephew, shouting for help, could not hold on. Barney plunged down the ship's side into the churning water. The alarms rang as Ship's Officer Clifford raced up, whipped off his coat and dived in to the rescue. The *Scot* stopped and a lifeboat was lowered to search for the two men hidden from sight beneath the heaving waves. After a long half-hour the lifeboat returned. Lying at the feet of the ship's doctor in the stern could be discerned the figures of two men. Along the crowds lining the rails flowed a murmur of relief: 'Barney Barnato is saved, thank God.' But as the lifeboat was hauled in it was seen that both men were unconscious. Under artificial respiration Clifford recovered, but after two hours of struggle there was no longer a flicker of life in Barney Barnato.

Even in death he influenced the stock markets of the world. For a millionaire to kill himself must surely mean his businesses were on the point of collapse. As news of the tragedy was telegraphed to the world's financial centres jittery stockbrokers marked down the stock in the Barnato empire and tried to offload it, only to have their fingers well-burned for their pessimism. As marks of respect, the Johannesburg Stock Exchange closed, the Cape Parliament adjourned, and work stopped for the day in all his mines. Barnet Isaacs, son of Isaac Isaacs of Whitechapel, who used to box in the evening in his father's little shop 30 years ago was Barney Barnato, millionaire twelve times over at 44. And dead.

CHAPTER ◆ TWELVE

Kimberley on the eve of War

With the publication of the damning telegrams at the trial of the 'Jameson Raiders' in Johannesburg, Cecil Rhodes' complicity in the plot was proved. But to the British public Rhodes remained a hero. Large contemporary cartoons showed him as either 'The Colossus of Africa' or 'The African Napoleon' with his feet based either on Cape Town and Cairo or on Cape Town and London. One popular illustration showed him relaxing confidently on board ship over the caption 'Returning to face the music'.

On 4th May 1896, *The Times* wrote: 'It would be idle to attempt to minimize the serious character of the cipher telegrams which were published last week', but it went on to dismiss suggestions that Rhodes should be tried as a criminal and that the charter of the British South Africa Company should be abrogated.

A correspondent calling himself 'Imperialist' vindicated Mr. Rhodes' career with warmth and eloquence, over two columns on another page beginning: 'The ablest Englishman in South Africa stands under a charge so grave ...' and ending with a summary of Rhodes' best aims. 'Imperialist' concluded: 'His objects have been the same as those which animated Drake and Raleigh and the great Englishmen who have made England what she is. If his methods have at times been such as might meet with approval rather from the statesmen of Elizabeth than of Victoria, it should with justice be remembered that the situation with which he has had to deal has been Elizabethan rather than Victorian.'

The Times' leader supported this line and noted: 'The project of a German-Boer alliance to cut off the British settlements in South Africa has been abandoned ... Mr. Rhodes has been the great obstacle to the accomplishment of these designs.' It admitted 'in his eagerness to counterwork them he has undoubtedly acted in a manner unworthy of a highly-placed and honoured servant of the Crown.' And concluded: 'His aims cannot be abandoned without disaster to the Empire. It is clear that the manifest destiny of South Africa is that all the European settlements south of the Zambesi shall be bound together in a federal Union.'

The enquiries by the Cape Parliament and British House of Commons acquitted Rhodes of involvement in Dr. Jameson's final act of invasion. Rhodes had in fact advised the doctor to wait, not on moral grounds, but because the time was not ripe. But his secret support of the rebellion up till the launching of the Raid was proved and condemned. Rhodes exhibited characteristic courage and candour by taking full responsibility for everything done in his name by subordinates.

His involvement however, now finally focussed the British Government's attention on its subjects' grievances in the distant Transvaal. It set out a fair and firm memorandum of the disadvantages and injustices its citizens were suffering. The Colonial Secretary invited Kruger to London for talks and the English Press promised the Afrikaner President a friendly welcome. President Kruger refused the invitation and the Colonial Secretary's recommendations. The situation was far worse than before. Quarrels between Boer and Briton in South Africa had widened into a rift. Rhodes, bitterly disappointed, resolved to leave his lovely home near Cape Town and to take up residence in Rhodesia. But his plans were delayed when the Matabele revolted in March 1896 and after several skirmishes, retreated to impregnable positions in the Matoppo Hills. Rhodes who had always been almost recklessly gallant – 'Jove! They're a bit close!' he once remarked coolly and unarmed in action – now resolved to go single-handed into the Matabeles' hideout. He moved his tent away from his troops into the foothills of the Matoppo range and waited there alone for 6 weeks while word of his presence was passed back to the African leaders. Then the chiefs invited him to a council in the heart of the hills where no troops could possibly have rescued him. With three friends, all unarmed, he followed the Matabele messenger back into the mountains. The chiefs outlined their grievances, Rhodes made and exacted some concessions and then asked 'Now, for the future, is it peace or war?' And the chiefs laying down their sticks as if they were surrendered arms, declared: 'We give you our word: it is peace.'

He had accomplished by faith and personality in a few weeks and without bloodshed what armies would have taken years to achieve. It was, in a single act, a mark of greatness. He continued to work for the federation of Southern Africa but war was very close now. The threat of it hung ponderously over Kimberley as Cecil Rhodes worked on at De Beers.

The journey up to the diamond city was still dirty, rough, and – as British travellers found – pretentious. 'There are no "hotels" in South Africa', wrote a visitor, 'but a number of shanties which if you picnic on any eatables which you may have with you and go to bed in an impenetrable bag, will serve you for a night.' 'Hotels' were shacks of bent iron at the side of the coach track, jumping with fleas.

Coaches supposed to stop for a night's rest only waited a few hours. Teams of 8 or 9 pairs were changed every hour or so at lonely sheds in the veld. A Boer was usually the whip, an African saw to the reins. A 'road' in Africa then was a track traced across the veld by parallel ruts from 3 to 30 yards wide.

Tickets for hotels marked '3s Good for 1 Bed' were handed out at Klerksdorp on the way up, but the tariff in the 'City of Diamonds' was usually 15s a day for full board. The English visitors sailed from London or Southampton to Cape Town, where 'on the magnificent steamers of the "Union" and "Castle" companies' as contemporary advertisements announced, 'passengers stepped ashore at Cape Town Docks without the use of tugs or small boats'. The steamers left England every week, arriving at Cape Town on Thursdays. Thence 'an express train leaves Cape Town within a few hours of the Mail Steamer, reaching Vryburg in 39 hours'. The railway rate for moving machinery and other general traffic up to Kimberley was 6s 6d per 100 lb.

The trains boasted 'Dining, sleeping and lavatory accommodation. Meals are served at Separate Tables at Reasonable Prices.' Time was also allowed at stations for 'Breakfast, Luncheon and Dinner'. The 774 miles by rail from Cape Town to Vryburg cost £8 1s 4d first class and £3 4s 8d third class. You could get from England right to Johannesburg in 22 days on the eve of the war, but this meant transferring to one of 'Gibson's favourably known "Red Star" Coaches' at Vryburg station, which delivered you in Johannesburg next day. Kimberley passengers stayed on the train, arriving at 8 a.m. next morning.

The traveller in diamonds probably used the Bank of Africa (established 1879: paid-up capital of £250,000) which from its City head office in Cannon Street, had branches in Kimberley and elsewhere. Leading London–South African merchants like Mosenthal Bros. and Co. had also established Kimberley branches.

Travellers were urged to try an 'Improved Double Roof Ridge Tent' made of 'Copper Rot-Proof Canvas' endorsed by the great explorer H. M. Stanley in his book *Darkest Africa*: 'I was well able to endure 200 days more of rain.' The London firm of Benjamin Edgington also supplied a 'Light simple Trestle Cot, Strong and Compact', together with mosquito curtains, india rubber bath and bucket and a waterproof sunshade, camp table and chair.

Westley Richards had been the name in guns and rifles since 1812. 'The reputation of our sporting rifles', they declared in 1899, 'for long-range shooting is unequalled and well established in all parts of South Africa.' Their 'Celebrated Falling Block Rifle' could be obtained for £20. Central Fire Double Guns cost 10 guineas.

If the traveller was going to engage in diamond mining himself he would use two famous suppliers: Commans & Co. of 52 Gracechurch Street, London, with their 'Otto's Aerial Ropeways – over 450 links at work – cheapest and best means of transport,' from whom he could also order diamond drills and air compressors for prospecting. From Ransomes, Sims and Jefferies Ltd., of Orwell Works, Ipswich came 'Portable Steam Engines, Mining Engines with Pit Gear and Winding and Pumping Engines.'

All this machinery was now profitably in action on the diamond fields. *The Pall Mall Gazette* – 'The most readable paper for English readers in South Africa – once purchased is not likely to be given up – Sixpence, each Thursday' reported a visit of inspection.

'The "blue ground", in which lie "the plums", is not very hard rock of a dull French-grey colour. The first thing, then, is to blast the "blue" with dynamite. This we can see done in the open workings. We are standing at the brink of Du Toit's. The warning bell has rung; Kaffirs and whites have streamed up into the Searching Room; here and there at the bottom of the great hole a puff of smoke, a spark of light, picks out the fuses dotted here and there over the floor. Every fuse has its appointed lifetime, and is lit in due order to allow the lighter a certain gauge for escape. Down in the yawning pit, a few men in charge take a last look and then flee to shelter. Presently, with deafening roar after roar, begins the fusillade. Masses of ground heave with a burst of smoke, tremble and tumble into gaps. Stones are thrown up almost to our feet. For ten minutes the great noise flaps and buffets round the chasm. Then another bell rings; the smoke clears away; and for twenty-four hours there is peace. Enough "blue" has been loosened for the next day's work.

'Now let us stand at the top of De Beers – at present the best organized of all the mines. What a change is here from the early days when the ground was hauled up by a Kaffir at a windlass or a horse at a "whim"! Here is an engine of, say, 1,500-Kaffir power, which has sometimes hauled out as much as 9,000 tons a day – a record unequalled anywhere else on or under earth, as I am proudly assured by Mr. Nicholl, the underground manager, who has come here from a crack Tyneside coal-mine. Up and down like a Jack-in-the-box pops the great skip, dashing 700 feet deep at every journey, to return with six tons of "blue", which at the top, with a prodigious somersault, it tips over into an attendant line of trucks.

'As I stand watching, suddenly there appears beside the skip – shot up out of the same shaft that the skip works in – a case like a very large double coffin, in which are packed, like figs in a box, four damp, hot, and soily figures. They emerge – a party of young English peers who, like myself, are visiting the diamond marvels. When my turn comes to be coffined, I lean back in the slanting shaft, taking care to protrude no hand or foot; a caution, a signal, then gentle motion, and the brilliant sunshine fades away. Once or twice on the way down my eyes are startled by a glimpse of dim-lit chambers with darkling figures mysteriously toiling, or my ears deafened by the rattle of the ponderous skip as it plunges up and down past the slower lift at headlong speed.

'At length I stand 700 feet beneath the ground, at the place where the skip is loaded for the ascent. All the passages of the mine converge upon a sort of oblong hell-mouth, tapering funnelwise to discharge into the skip below. The jaws of this are four trucks wide, four trucks going to a load. Here stand four herculean shapes, and as the stream of full trucks from the various tramways reaches them, these four seize each a truck, force it against the lip of the hole, and all together, with a shout, upset the weighty convoy. Instantly they drag back the empty trucks, to be pushed away each by its own Kaffir for refilling in the dark and sloppy labyrinths. A sign, meanwhile, has throbbed to the engine-room above, and almost before it has touched the bottom the skip, with its 6 tons on board, is on its upward race again.

At the surface the precious "blue" is run in trucks by an endless rope to the drying-grounds, which are some miles away and some square miles in extent. Each truck-load – 16 cubic feet, or about a ton of "blue" – conceals on an average a carat and a quarter of diamond, ranging in value from 3s 6d to £20 a carat. On the "grounds" the "blue" is softened by the sun and air, broken with picks, and then conveyed back to begin that process of reduction, which magically transmutes each ton or two of dull, heavy earth into a tiny brilliant.

'Kimberley, straggling in brick, and clay, and iron, stands in the level veld. For miles around there is nothing much higher than the dwarf bush. Except around and amid the City of Diamonds itself there rise certain low, sleek, drab, uncanny-looking hills of dirt. These are not hills, but "tailings".'

The contemporary magazine finally describes the washing method of screening diamonds, which has not noticeably altered in the last 70 years. The 'pulsators' chug on today.

'First, the ground goes into the washing machine – the primitive 'cradle' on a large and perfected scale – the working of which depends on the fact that the high specific gravity of the diamond makes it behave differently from other stones under the joint action of centrifugal force and gravitation. Spun round in perforated cylinders and pans under a whirlpool of water, the bulk of the ground flows off in "tailings" of grey mud. The residue of divers stones of divers sorts and sizes is then jogged about with more water in the "pulsator". This machine is a huge framework of graduated sieves and runlets, which sorts the stones into several sizes, and after much percolation delivers each uniform lot at a separate receptacle. After the pulsator, there remain a number of "dry-sortings" and re-sortings on various tables, by hands both black and white, all under lynx-eyed surveillance, the pretty red garnets and other valueless pebbles being swept off by dozens with a bit of tin, the diamonds dropped into a sort of locked poor-box; until finally the coveted hoard, all scrutinized, classified, and valued, lies on the office table of the company on its way to their impregnable safes.'

CHAPTER ◆ THIRTEEN

War, Siege, and the end of Rhodes

The South African War, which had seemed inevitable for years, finally broke out on 11th December 1899. The conflicting attitudes and interests of the British Colonies and of the two Boer Republics (particularly of the Transvaal) had grated against each other for 60 years.

The conflict between a small struggling new nation and the most powerful Empire of the nineteenth century came to a sharp head and a bitter long-drawn ending. Good and bad on both sides became entangled: Boer myopic stubbornness coupled with a proud courage; and British ignorance and inconstancy coupled with a final generosity.

Who had in any case the original 'rights' to the country?

The first permanent white settlement north of the Vaal River had been made by a party of Boers under Potgieter in 1838. These families set up a form of government and then entered into a confederation with Boers settled in Natal and south of the Vaal. By 1852 these Boers at Potchefstroom were led by Commandant Andries Pretorius. At a meeting with him at a farm on Sands River the British (acting through assistant commissioners, instigated by the High Commissioner) formally recognized the independence of the Transvaal Boers who, after 5,000 families had trekked into the new country, now numbered nearly 40,000. The first clause of the Sands River Convention, as it came to be called, set out the full British recognition of this settlement clearly and generously.*

*'The assistant commissioners guarantee in the fullest manner, on the part of the British Government, to the emigrant farmers beyond the Vaal river, the right to manage their own affairs, and to govern themselves according to their own laws, without any interference on the part of the British Government, and that no encroachment shall be made by the said government on the territory beyond to the north of the Vaal river, with the further assurance that the warmest wish of the

In 1856 an Assembly at Potchefstroom created 'The South African Republic' with Marthinus Pretorius as its first President. Some rebellious and outlying groups of settlers had to be forced into agreement, and in order to strengthen central government Pretorius attempted, first by negotiation and then by force, to persuade the new neighbouring Orange Free State into a Union with them. Words failed and the Transvaalers raised a Commando. But the former Orange River Colony remained independent and in the Transvaal the Africans rose and the Republic tottered towards bankruptcy on a total annual revenue of £31,511. By 1877 one shilling would buy a Transvaal note purporting to be worth 20 shillings.

Two external pressures now squeezed the little bankrupt state. The Zulus were menacing its southern border, and in Cape Colony and London there were those plans to bring all southern Africa into a confederation under British control. The perilous state of the Transvaal seemed to need urgent support, so Sir Theophilus Shepstone was instructed by Britain's Colonial Secretary to visit the Transvaal and to annex it to Britain, if he and its inhabitants thought fit.

Shepstone rode into Pretoria with 25 mounted policemen in January 1877, to confer with the Raad (the Boer House of Representatives). Although the Transvaal Government owed £215,000 to the public, could not pay its contractors' bills, and was about to be attacked by Cetywayo, the Zulu chief, the Raad would not agree to Shepstone's proposals for federation. Shepstone therefore by proclamation and without bloodshed promptly annexed the Republic.

The majority of Boers still wanted their independence. They generally gave Britain no aid in her war against the Zulus, and when British forces were massacred by Zulu Impis at the disaster of Isandhlwana, the Boers redoubled their agitation to be free of the British flag.

They continued to pay no tax. Where their own people had failed to collect, the British collectors had no chance at all. Then an armed Boer commando recaptured a farmer's waggon which had been

---

British Government is to promote peace, free trade, and friendly intercourse with the emigrant farmers now inhabiting, or who hereafter may inhabit, that country; it being understood that this system of non-interference is binding upon both parties.'

destrained by the local sheriff in lieu of tax, and in December 1880, the Boers declared themselves a republic again, and attacked British forces all across the country.

Worse was to come for Britain. In January 1881, the Boers invaded Natal, defeated British troops under Sir George Collet in two initial engagements and then overwhelmed his force at Majuba Hill.

This major reverse reverberated for years, but London's attitude was that of an absentee landowner whose bailiff had been badly stung by hornets: the place would be better left alone. The Boers' victories over regular British troops however made them confident that they could always defeat the 'Imperialists' and ensured their desire to put the issue to the sword eighteen years later. After Majuba Hill the British Government promptly came to terms with the Transvaal, acceding to almost all the Boer demands and granting the Republic complete internal self-government. British settlers in Pretoria were so disgusted by this sell-out that they dragged the Union Jack through the dusty streets of the Transvaal's little capital.

With this background it was impossible that Boers and English in Johannesburg should live quietly together. The repression of the Rand's Uitlanders under President Kruger exacerbated an already prickling relationship. When Barberton sprang into existence as a goldmining town of 5,000 inhabitants the London Convention* established fair conditions for foreigners in the Transvaal. But the conditions were not kept.

Kruger's refusal to give a vote to the 'Uitlanders', his maintenance of monopolies on most essentials including the vital gunpowder and dynamite and his imposition of heavy taxes, led to the formation of the Reform Committee and the abortive Jameson Raid. The situation naturally did not improve thereafter. Petitions to the Raad by the 'Uitlanders' were jeered at and thrown out. In 1898, 21,000 'Uitlanders' finally resolved to invite official British intervention. They dispatched a telegram to Queen Victoria:

*Its Article 14 stated:

'All persons, other than natives, conforming themselves to the laws of the South African Republic (a) will have full liberty, with their families, to enter, travel, or reside in any part of the South African Republic; (b) they will be entitled to hire or possess houses, manufactories, warehouses, shops and premises; (c) they may carry on their commerce either in person or by any agents whom they may think fit to employ; (d) they will not be subject, in respect of their persons or property, or in respect of their commerce or industry, to any taxes, whether general or local, other than those which are or may be imposed upon citizens of the said Republic.'

'The condition of your Majesty's subjects in this State has become well-nigh intolerable. The acknowledged and admitted grievances, of which your Majesty's subjects complained prior to 1895, not only are not redressed, but exist today in an aggravated form. They are still deprived of all political rights, they are denied any voice in the government of the country, they are taxed far above the requirements of the country, the revenue of which is misapplied and devoted to objects which keep alive a continuous and well-founded feeling of irritation, without in any way advancing the general interest of the state. Maladministration and peculation of public moneys go hand in hand, without any vigorous measures being adopted to put a stop to the scandal. The education of Uitlander children is made subject to impossible conditions. The police afford no adequate protection to the lives and property of the inhabitants of Johannesburg; they are rather a source of danger to the peace and safety of the Uitlander population.'

Sir Alfred Milner in the Cape forthrightly backed the 'Uitlanders' in a strong cable to London describing the British citizens in the Transvaal as 'helots', and avowing that the case for British intervention was 'overwhelming'. It was not lost on Whitehall's classically educated heads that a 'helot' was a class of serf in ancient Sparta deliberately humiliated and liable to massacre ...

A meeting in May 1899, between Milner and Kruger at Bloemfontein in the Orange Free State bore no fruit: Kruger would grant no concessions to the 'Uitlanders'. Pro-British supporters in South Africa hardened their resolve: no solution would do unless the 'Uitlanders' had equal rights.

War's collision course was now set on both sides. But while the British only faintly expected they might have to fight again, the Boers were already resolved; they procrastinated only so that by October the grass would be growing on the veld. In that spring month, the Transvaal at last achieved its alliance with the Orange Free State. An ultimatum was presented to the British Agent in Pretoria. It fearlessly – or impertinently – demanded the instant withdrawal of all British troops from all Boer frontiers and the return to Britain of all reinforcements just landed at the Cape or en route for South Africa.

The ultimatum caused some apoplexy at the Boers' impertinence, but it stiffened Britain's resolve to fight. She refused the terms 48 hours later. At once a Boer commando attacked a British armoured train south of Mafeking and a Boer Army encircled Kimberley, trapping Cecil Rhodes.

The bold Boer plan was to invade Natal. Striking south and east they inflicted several defeats on the British and laid siege to Ladysmith. Simultaneously they struck westwards to encircle Mafeking with a Transvaal force under Cronje, while Orange Free State commandos reinforced the siege of the city of diamonds.

At Cape Town Britain landed an Army Corps of 3 Divisions under General Sir Redvers Buller. But they could not invade the Free State until they could free the defence forces pinned down in Natal and cooped up in the three besieged cities. They particularly wanted to liberate Kimberley. For political and economic reasons they needed to rescue Cecil Rhodes and the diamond mines, and the incarcerated 'Colossus' was sending a spate of signals urging no delay.

The besieging forces had arrived on the town's outskirts on 14th October 1899, and by next day, as the mines' hooters screeched their warnings, had fully encircled Kimberley.

The defence consisted of the North Lancashire Regiment, the Kimberley Regiment, and the Town Guard, a sort of civilian militia which included, naturally, a company raised by members of the Kimberley Club and called 'The Buffs'. This was stationed under Captain Mandy at the Belgravia Redoubt, while the overall military commander, Lt. Colonel Kekewich, sensibly made his headquarters in the gentlemanly atmosphere of the Club.

Cecil Rhodes however did not. The former statesman dismissed any notion of being under the orders of a mere Army colonel and removed himself to a base of his own in the Sanatorium whence, as head of De Beers, he set about organizing the town's defences.

The mines went on producing diamonds for a couple of months but were closed down completely when it was realized that the war was going to be a long grim business.

As the weeks passed and shelling of the corrugated iron buildings grew heavier, the diamond mines were thrown open as shelters against the Boer bombardment. The besieging forces possessed a damaging gun called 'Long Tom', which was particularly directed at the Club and the Sanatorium. Both remained unscathed but several injured members were treated for wounds in their own premises.

The city was suffering badly. In the Kimberley Engineering Works

Rhodes organized a counterblast. George Labram, who was later killed when the Grand Hotel was hit, designed a huge gun, named 'Long Cecil' to fire home-made shells exploded by De Beers mining dynamite. Some siege incidents smacked of an annual office outing, as when the first shell of 'Long Cecil' was ceremoniously fired at the besieging Boers by the wife of a Kimberley Club member.

De Beers workshops also constructed an armoured train and installed a cold storage compound for 200 cattle which had been rounded up and pirated just as the siege began. As food shortages gripped the town, ration cards were issued even for horse meat. There was, however, a cache of delicious food in the American Stores in Dutoitspan Road, and this was evidently the provider for the Kimberley Club's Christmas Dinner for which the menu read:

KIMBERLEY CLUB

XMAS DAY – SIEGE DINNER
1899

Anchovies Olives
Turtle Soup
Mutton Cutlets and Peas
Aspic of Foie Gras
Roast Turkey and Ham
Asparagus
Boiled Potatoes
Plum Pudding
Preserved Ginger

Stilton Cheese Tomatoes
Dessert Coffee

The menu seems incredible and perhaps it was: some current members of the club believe it was a joke in good taste.

Rhodes continued to issue his own signals to the outside world to the chagrin of Colonel Kekewich, who had the misfortune to lose his Chief of Staff in a skirmish on Carter's Ridge soon after the siege began.

From Cape Town Buller sent one Division under Lord Methuen to relieve Kimberley. Methuen began his march in November, drove the Boers out of Belmont on the 23rd and pressed on to the Modder River. Here the line was held by the expert commando leader De la Rey, reinforced by 2,000 men under Cronje who had arrived from Mafeking the previous night.

The British vainly assaulted the river line all day. Then, during the night, De la Rey's Boers fell back from their bank. But Methuen had lost momentum. Despite further demands from Rhodes and the need to liberate Kimberley, the British Division had to pause and consolidate on the Modder.

The size of the task had now dawned on the astonished British at home. 'The band of Afrikaaner farmers' had successfully invaded the British colonies. Three main British colonial towns, Kimberley, Ladysmith and Mafeking, were still besieged. A former colonial Prime Minister was still holed up. The gold and diamond industries had ceased to function. Lord Methuen had been wounded, and was halted on the Modder river. At home, no less than 5 Divisions had already been called up to try to win the war and now a sixth was raised. The offer of other colonial troops was eagerly accepted and recruiting stations opened at South African ports. The home resources of the world's most powerful Empire had been emptied by the requirements of this tedious war in southern Africa. At home white feathers flew as girls goaded their men into volunteering to fight.

On 11th December Methuen attacked Cronje who was holding the Spytfontein and Magersfontein kopjes between the Upper Modder River and the Kimberley Road. The battle ensued in diamond country and resulted in the slaughter of the Highland Brigade which, attacked by the Boers while still forming up for a night assault on Magersfontein, lost its General and 750 men. Next day Methuen's remaining two brigades failed to restore the position.

Over in Natal the war went equally well for the Boers and in Cape Colony their two commanders of genius, De la Rey and De Wet, were carrying almost all before them in brilliant, fluent sweeps across country. The Boers were well-mounted, knew the country well and blended into it. Against the British infantry slowly proceeding in obvious regimented columns, the Boer cavalry attacked like hawks on hampered hound-dogs. Adversity however breeds good generals. At last there emerged an equally expert adversary in General French with his 2 Cavalry Brigades.

On 10th January 1900 British hopes were raised by the arrival at Cape Town of Lord Roberts and his reinforcements, but sank again only a fortnight later as their forces in Natal suffered the crushing defeat of Spion Kop and lost 1,700 men.

Lord Roberts planned to withdraw General French's cavalry from Colesburg for the relief of Kimberley, where disputes between Rhodes and the military authorities had now reached the height of bitter fury. The town was under damaging shellfire and short of food. But Cronje still sat firm on Magersfontein. General French swung eastwards to avoid Cronje's army, skirmished with De Wet, and pushed his 2 brigades across the Modder at Klipdrift. He was now within strike of Kimberley and had time to send a message into it by heliograph saying that when he entered the city Mr. Rhodes must leave it! Cronje, mistaking the main British plan, sent too few troops to oppose French and the British cavalry, sword in hand, broke up the Boers and swept through to relieve Kimberley on 15th February.

Among French's column riding into the town down cheering streets and under tossing flags were the future Duke of Athlone, who would return to a united South Africa as Governor-General, and the future Field-Marshal Allenby, then a cavalry major. General French would end up Field-Marshal the Earl of Ypres, for he was to become Commander-in-Chief of the British Expeditionary Force in France in 1914.

General French and his entourage moved briskly into the Club to be refreshed by £20 worth of champagne. Cecil Rhodes' health had deteriorated quickly during the long siege, and he now discreetly seized the opportunity to move out and travel south to his Cape Town home.

French's cavalry were soon in action again, leaving Kimberley for the north-west to head off Cronje, whose rearguard, as he attempted to slip away westwards, was being harassed by Kitchener. At Paarderberg on 29th February Cronje's force of 4,000 men was caught between Kitchener and French and compelled to surrender. It was the first major British victory of the war.

As Cronje surrendered, news came in from the east of the relief of Ladysmith. The Boer forces in Natal and Cape Colony started to withdraw northward and the Presidents of the two republics put out peace feelers. They were not, however, giving any political ground away. They wanted a return to the status quo at the outbreak of fighting and this Britain could not contemplate.

Bloemfontein was captured by Lord Roberts on 13th March and Johannesburg was seized on 31st May, President Kruger fleeing to the sea with the State papers on the eve of the British occupation.

After Lord Roberts captured Pretoria on 5th June, official resistance formally ended and the war seemed to be over, but the last and most bitter period was now to begin. For the next two years courageous and brilliantly-led Boer commandos harried the British occupying forces across both republics and sparked risings across Cape Colony. The British, unpractised in coping with partisans, blundered about trying to catch De Wet's flying phantoms. Hertzog and his guerillas raised the rebellion in the Cape and even reached the Atlantic and fired on a British warship.

As the flames of revolt spread, the raiding parties grew from little groups to big battalions. Botha in the Transvaal had no less than 1,900 commandos with which to attack British garrisons and railways. Every week there were further Boer successes and the less mobile British were for ever hunting the horse-tails of the partisans who struck, then vanished into the veld.

By Christmas 1900 with the war officially 6 months dead, Lord Kitchener was demanding more reinforcements, and a further 30,000 cavalry were raised in Britain and sent out. Kitchener now resolved to sweep great areas of the Republics quite bare, using his cavalry to denude the country of crops, livestock and people. Non-combatants were locked in concentration camps where disease took a frightful toll. The hatred for Britain which still smoulders in some South African hearts is fuelled by memories of the bad days of the concentration camps, Kitchener's scorched-earth policy, and his 'drives' across country between the block houses.

Finally the Boer leaders were convinced that further fighting was futile. Britain had lost 5,700 killed and 22,000 wounded. The Boers' casualties were less, but in March 1902, as their leaders rode into Pretoria to treat, they had lost 40,000 prisoners, in British hands.

On 26th March, Cecil John Rhodes, the colossus of Africa, died at the Cape. He had finally fallen ill not in the great mansion Groote Schuur, which he left to the nation, but in his little thatched seaside cottage at Muizenberg, where the Indian Ocean rolls warmly in against Cape Point's eastern flank and the surfers swoop on the waves.

What he stood for still lives on, for he left in his astonishing Will virtually all his gigantic fortune to finance the Rhodes Scholarships at the Oxford he loved, for citizens of the former British colonies and of the United States.

The new King Edward VII was represented at Cecil Rhodes' memorial service at St. Paul's. Ten members of the Cabinet attended, so did ex-Ministers like Asquith, the Commander-in-Chief Lord Roberts, the American and German Ambassadors, Members of Parliament and peers of the realm, the Directors of De Beers and the British South Africa Company, the Vice-Chancellor of Oxford University, the representatives of Queen Alexandra and the Prince of Wales, the Lord Mayor of London and the Sheriffs, Pierpont Morgan the American millionaire, General Booth of the Salvation Army, and thousands of unknown citizens who filled every foot of the Cathedral an hour before the solemn service started.

The band of the Coldstream Guards played, the brass rang out under Sir Christopher Wren's dome and the booming drumbeat of Handel's *Dead March from Saul* rolled across the bowed heads in the nave.

Five thousand miles from the Empire's chief city another ceremony was taking place. Rifle shots cracked in the Matoppo Hills of Rhodesia as the great man's body was lowered into its lonely grave at World's View. Further south still in Johannesburg, all businesses closed and flags flew at half-mast.

Rhodes' body had been transported north in the special De Beers Pullman car which he had ordered from America for the company's board meetings. It had thus come up to Bulawayo by the rail he planned, and then been taken through the town on a gun carriage behind men of his BSA Company's Mounted Police. It was escorted by all the local dignitaries, dressed not in the civic black and top hats of London, but in the breeches, boots, open shirts and slouch hats of frontier pioneers. From the outskirts of the town Rhodes' body was taken up into the lonely heart of the mountains and after it came all the local population, including thousands of Matabele. As the coffin was lowered into the granite vault Bishop Gaul read Kipling's poem, and Sekombo, the great Matabele orator, declared that Rhodes' spirit was now with the spirit of Umsiligazi, founder of the Matabele nation.

It was over. To the south in the Transvaal, Generals Botha and De Wet arrived in Pretoria under 'safe conducts' to start peace talks. Skirmishing was reported and prisoners were still being taken in the Orange River Colony and the Western Transvaal. A mounted contingent of the Durban Light Infantry were seen off to the front by Natal's Prime Minister, and two members of the Canadian Mounted Rifles were shot by the Boers after an encounter at Klein Hart's River. In the Cape near Carnarvon and Sutherland Boer raiders still harried British forces. But the war was dying too. From annexation the two former republics moved smoothly towards self-government, and then to federation. The dream of Cecil Rhodes was finally fulfilled. On 31st May, 1910, the Union of South Africa was established.

The Rhodes Scholarships: Excerpt from Cecil Rhodes' Will

1. Colonial. – 'I consider that the education of young colonists at one of the universities in the United Kingdom is of great advantage to them for giving breadth to their views, for their instruction in life and manners, and for instilling into their minds the advantage to the colonies as well as to the United Kingdom of the retention of the unity of the empire.'

2. American. – 'I also desire to encourage and foster an appreciation of the advantages which I implicitly believe will result from the union of the English-speaking people throughout the world, and to encourage in the students from the United States of North America who will benefit from the American scholarships to be established for the reason above given at the university of Oxford under this my will an attachment to the country from which they have sprung, but without, I hope, withdrawing them or their sympathies from the land of their adoption or birth.

'My desire being that the students who shall be elected to the scholarships shall not be merely bookworms, I direct that in the election of a student to a scholarship regard shall be had to (1) his literary and scholastic attainments; (2) his fondness for and success in manly outdoor sports such as cricket, football and the like; (3) his qualities of manhood, truth, courage, devotion to duty, sympathy for, and protection of the weak, kindliness, unselfishness and fellowship;

Above: Colesberg Kopje, 1871. First prospectors on land which became part of Kimberley's 'Big Hole'. Below: Kimberley Mine 'Big Hole', 1873. The 30 foot square claims going down.

Above: Kimberley Township, c. 1876. *In the foreground: shacks of early diamond merchants. Below: Kimberley Mine,* 1877. *Now ropeways haul up buckets of diamond-bearing ground.*

Below: Kimberley's 'Big Hole' as it is today. One mile in perimeter, 1,300 *foot deep. It 'died' when war started on August* 4, 1914.

Above: The Central Company's Kimberley Mine Shaft. Smaller claims had now been amalgamated or been bought out.

Above: Barney Barnato 1852–1897. *Arrived in Kimberley in* 1873, *bought his first claim in* 1876, *floated his first company in* 1881, *became a fourteenfold millionaire and died a suicide in* 1897.

Above: The Rt. Hon. Cecil Rhodes 1853–1902. *Colonial Prime Minister, pioneer of diamond market control, founder of De Beers and creator of a country.*

Above: 'The Cullinan', 1905. *The largest diamond ever found weighing* 3,106 *carats in the rough, as shown.*

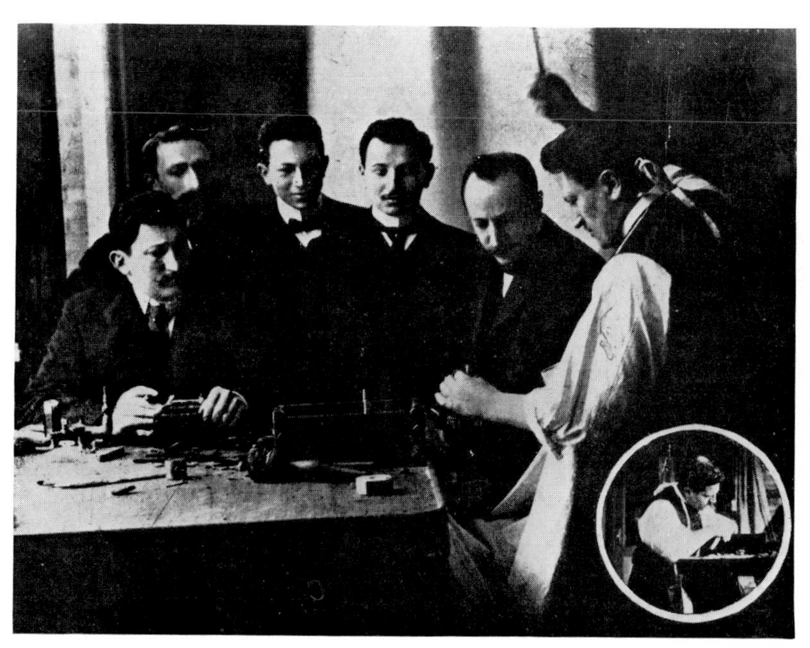

Above: 'The Cullinan' about to be cleaved by Mr Joseph Asscher of Amsterdam on February 10, 1908.

Top left: 'The Cullinan' after cleaving.

Above: Nine major stones cut from the Cullinan.

Left: The magnificent 'Cullinan 1' known as the 'Star of Africa'.

Above: The Finsch Mine, and recovery plant. Discovered by Allister Fincham and his partner prospecting for asbestos in 1960. *It was the first diamond pipe to be opened in South Africa since Premier in* 1902.
Below: Continuous Grease Belt. Finsch Mine recovery plant. Water repellent diamonds adhere to grease while other materials are washed away.

Above: Rough Gem Embedded in Blueground (Kimberlite).

Top Left:

Diamond being marked to show sawing direction. It is held in a 'dop' against a rapidly revolving phosphor-bronze disc impregnated with diamond dust.

Centre Left:

Stone about to be cleaved along its grain by a wedge-shaped knife.

Bottom Left:

Diamonds being polished.

Overpage:

Bruting a diamond to give its shape. Held in the revolving lathe while another diamond 'dop' is brought to bear against it.

Overpage:

De Beers Synthetic Diamond Dust magnified forty times. Synthetic diamond abrasive is used in saws for cutting stone, concrete, etc.

Right:

Elizabeth Taylor wearing her 69.42 carat diamond.

Below:

Agnes Sorel. The 15th century French lady who was the first woman commoner to wear diamonds.

and (4) his exhibition during school days of moral force of character and of instincts to lead and to take an interest in his schoolmates, for those latter attributes will be likely in after life to guide him to esteem the performance of public duties as his highest aim.'

CHAPTER ◆ FOURTEEN

The Greatest Rock of All

The war had hardly ended before two events occurred of prime importance to the diamond industry. In 1902 diamonds were found for the first time in startling numbers far from Kimberley. And in the same year a young man named Ernest Oppenheimer arrived in Kimberley from London. The two events were unconnected at the time. But the new Premier Diamond Company formed near Pretoria and the new young arrival were going to come together in the future with remarkable effect.

The discovery of diamonds in one isolated 'pipe' 300 miles north-east of the recognized and established Kimberley area freshly illustrated diamonds' rule of chance. But the strike was a matter of instinct based on experience. The man whose instinct led him to the farm Elandsfontein 24 miles east of Pretoria, was Percival White Tracey who had worked a claim in De Beers' original mine at Kimberley 20 years earlier and who was a former acquaintance of Cecil Rhodes. Following the discovery of gold on the Witwatersrand in 1886, Tracey, like many diamond prospectors who had done well but not brilliantly, decided to move on and try for a gold strike on the Rand. He moved to Johannesburg, was successfully involved in the Simmer and Jack goldmine, but continued prospecting outside. Pretoria was a short haul from the City of Gold even in Tracey's dog-cart days. (You drive there now in minutes over modern roads and high rolling land through the swift spread of suburbia which almost bricks the two cities together.)

On one of Tracey's trips he came across traces of diamondiferous soil in the bed of a stream. He recognized it immediately from his Kimberley experiences, and reckoned that the soil had probably been washed down by the stream from a 'pipe' of kimberlite which had once erupted somewhere in the neighbourhood. Very well: he would trace back the stream's course. He paddled along the stream like a hound and was thrilled to discover in the right place a small low hill closely resembling that richly fruitful 'kopje' at Kimberley. This could be it. He pressed forward.

But his quest was instantly obstructed. The farm was owned by a farmer who was not merely stubborn: he was irate and often menacing. Joachim Prinsloo had twice had to sell his farms to damn gold prospectors. Certainly he had made prodigious profits both times, but he hated strangers; he loathed moving. This time he was settled for good.

His original land had been right in the middle of the 1886 Madderfontein gold strike. Very reluctantly he had agreed to sell it and with part of the proceeds bought another farm further away at Kaalfontein, cursing those grasping prospectors all the way to the bank. He settled in. He settled down. But once again Mr. Joachim Prinsloo was sitting literally on a gold mine. Once again speculators, prospectors, agents, spies and touts, drove up daily to his door pestering him to sell, offering him crazy amounts so that they could move him out, dig up his decent farmland, and extract mountains of earth for a few traces of ore – a lunatic business. But in the end, for the sake of peace, he sold again and moved further on. He was now entrenched on Elandsfontein for the rest of his days. But he had been twice bitten. He sat on his 'stoep' looking over his land. He was on guard against prospectors: by his chair stood a loaded rifle. At this point he suddenly and to his fury observed Tracey attempting to dig on his land. Tracey asked him the now nauseous question: how much would he take to sell and move out? The farm was not for sale. Ever! said Mr. Prinsloo. When Tracey called again in his Cape cart Prinsloo shouted that if he got out to set foot on his land again he'd shoot him. Prinsloo's fury and his rifle barrel were so menacing that Tracey withdrew to consider his next move.

He resolved to team up with a leading Johannesburg building contractor, Thomas Cullinan, who had made his fortune as the brand-new city went leaping up. Cullinan had adequate funds, was experienced in property speculation and skilled in negotiation. By some deception Cullinan and Tracey succeeded in getting onto the farm again unscathed, probably (the story was denied) by pretending to be cattle inspectors making an agricultural check. This time they definitely established the presence of diamonds. The skilled Cullinan opened silver-tongued negotiations. All to no avail. The years passed. Prinsloo remained obstinate. Tracey and Cullinan remained eager. Like wolves outside a barred sheep-fold, they sat bright-eyed, waiting and finally old Prinsloo died. The farm passed to his less recalcitrant daughter. They were able at last to negotiate a sale. Bringing in a third partner, a Mr. Jerome, to help with the considerable finance needed they bought Elandsfontein farm for the huge price of £52,000. The Premier (Transvaal) Diamond Mining Co. was founded in 1903 with Cullinan as chairman and Tracey as managing director. Mining began in April. The kimberlite was hauled by endless ropes from the open-cast pit up to the washing-pans and a pulsator was installed to speed the sorting of the diamonds from the sludge. It was in full blast early, and the hauls were dazzling rich.

Within two years even these lucrative shoals were surpassed: on 25th January 1905, the largest diamond the world has so far seen was spotted by Mr. F. G. S. Wells, the surface manager. It was so close to the surface in the side of the pit that Wells simply hopped down to dig it out. He must nearly have fainted at the sight of it, mistrusting his eyes: the uncut rock could hardly be grasped in his hand. It measured 4 inches by 2½ inches by 2 inches and weighed the unbelievable amount of 3,024¼ carats. This dwarfed the two huge stones produced by Jagersfontein the 'Excelsior' (969½ carats) and the 'Jubilee' (634 carats). After the initial celebrations and the presentation of a cheque for £2,000 to the blessedly keen-eyed manager, a reaction took place in the diamond-mining world. How, it was asked, could one dispose of such a vast stone which was literally invaluable? After all the 'Excelsior' had been 10 years with the Kimberley Diamond Syndicate waiting for a buyer. The second problem concerned a chronic anxiety; the Premier had in its first 2 years of working disgorged 22 stones each larger than 100 carats. How was the general marketing of Premier diamonds to be controlled to safeguard South African prices everywhere?

The first question was adroitly answered. The huge stone named the 'Cullinan' after Premier's chairman was bought by the Transvaal Government for £150,000 and presented to Edward VII as a birthday present from colonial taxpayers to their Monarch. The King sent it to be cut by Asscher of Amsterdam who, after lengthy cogitation, suggested that he try to divide it into 9 major stones, 96 smaller brilliants and a number of polished ends.

To foil thieves at the start, the huge uncut stone was posted unregistered by ordinary mail from South Africa to London. From London it travelled to Amsterdam in Joseph Asscher's pocket. There Asscher and his brothers studied the stone for days debating, as all diamond cutters must, how best to cleave. They were sharply aware that this time the eyes not only of the experts but of all the sophisticated world were watching for one false tap of that hammer. At last they presented their plan to the King, who approved it. At last Asscher stood, mallet poised, looking down for the thousandth – and last time – on the gigantic stone. Finally he struck. The diamond split exactly as intended. In the released gasp of great relief all round, Joseph Asscher tottered. He started to swoon and a doctor came racing to his rescue.

Cutting proceeded perfectly to plan. All the nine major stones were incorporated in the Crown Jewels, and the largest, weighing 530 carats and named 'The Star of Africa', was mounted in the Royal Sceptre.

Tom Cullinan, builder, became Sir Thomas, and the early excavations at Premier started to develop into South Africa's most modern diamond mine. Something else possibly lurks in the future. The shape of the original 'Cullinan', particularly of its one large face, has suggested to many experts that it was once just part of an even more colossal stone, split from it by subterranean pressures. If so, does the monster lie deep down beneath old Prinsloo's farm? Or has it already been smashed into fragments by the teeth of recovery machinery geared for mundane diamonds?

The second more general problem accentuated by the 'Cullinan's' discovery was going to take longer to solve. But the man to solve it (among much else in southern Africa whose industrial emperor he was to become) was the young Ernest Oppenheimer who had just arrived in Kimberley as a diamond dealer. In the Diamond City the issuing of licences to dealing firms like his had done much to restrain Illicit Diamond Buying. There had been a natural increase in the number of firms dealing and broking in diamonds as soon as it became illegal to traffic in diamonds without a licence. As legislation grew more complex, more skilled men were needed in this new profession. They needed an eye for diamonds, a quick mind for figures, substantial capital backing in London, and a shrewd knowledge of the potential markets. Often diamond-dealing firms were off-shoots of big merchants like Mosenthals, and it was as a representative of Mosenthals that a Mr. Anton Dunkelsbuhler had arrived on the diamond fields in 1872. The company which later bore his name was London-based and established in Hatton Garden.

Anton Dunkelsbuhler was very popular and highly regarded around Kimberley. His firm became equally successful. It joined the Diamond Syndicate which handled the output of most of Kimberley's major diamond producers. The Syndicate was already a vertical monopoly from production to marketing through those of its members who were additionally members of diamond-producing partnerships. The Syndicate's links with De Beers were already strong, but they were not yet entwined.

The need for a Syndicate had been emphasized in Kimberley by a comparison of the diamond industry with that of the gold producers. Kimberley was in a far more precarious position than Johannesburg: gold was a mineral which commanded an insatiable industrial market, diamonds did not.

When young Oppenheimer started working for Dunkelsbuhler in London there was only a very small demand for industrial diamonds and the market for gem stones was as fickle as it is today. The birth rate, the standard of living, financial booms and slumps, variations in public taste all blew sharply and in different directions upon the demand for engagement-rings and jewellery. The marketing of gem stones had to be trimmed to these changing winds. This rôle the Diamond Syndicate was attempting to fill.

Ernest Oppenheimer was born in Germany in 1880, the eighth child of German parents who had a wide web of relations. He started with Dunkelsbuhlers in 1896 as a junior clerk sorting diamonds on long tables. He was tenuously related to Anton Dunkelsbuhler and he had brothers and cousins in the firm. His brother Louis, with whom he remained particularly friendly all his life, was made a partner in 1901. In that year Ernest came of age and immediately took out British citizenship. In 1902, the year of Rhodes' death and the signing of the Peace of Vereeniging, he was posted to Kimberley as Dunkelsbuhlers' representative. Diamond output that year, restrained by the war, was worth just under £5 million.

The Syndicate, with which De Beers had just concluded a profit-sharing agreement, comprised 8 firms of which Dunkelsbuhlers held 12%. De Beers, until the discovery of the Premier Mine, dominated

South Africa's diamond production, as it dominated the town of Kimberley. It did not own any alluvial diggings and it owned only some minor interests in the outlying mines. But its complete control over the five central mines' production and its links with the Syndicate which controlled sales gave it a giant's strength.

The Premier was different. Not only did it lie miles beyond the Kimberley aegis, but in 1903 60% of it was owned by the government. Both the government and the minority shareholders wanted maximum production and maximum sales. They rejected any sort of control from Kimberley and wished that place to the devil if it couldn't compete without restrictions. Alfred Beit, whose firm of Wernher Beit was an important member of the Syndicate, visited the Premier Mine to check what was happening: he literally had a stroke brought on by what he saw. Premier production was soaring uncontrolled. Beit begged De Beers to come to some agreement with their rivals: Premier, he complained, had leapt from 750,000 carats in 1904 to nearly 1,900,000 carats in 1907. This almost equalled the total De Beers output which remained at just over 2,000,000 carats. What was worse, the Premier's diamonds were by-passing the Syndicate and were being marketed independently through its own London sales office.

Then at the end of 1907, another slump gripped America and Europe. Demand dried up, sales snailed and the Syndicate was suddenly in difficulties. It had agreed in October to buy every month £450,000 of diamonds from De Beers and £193,000 from the Premier. Everything then had looked set for cosy co-operation. Now in the slump of 1908 the Syndicate was stuck with a stock of £3 million worth of diamonds on its hands which it could not sell. It had to go back on its agreements and refuse to buy any more. It offered to 'freeze' its accumulated stock releasing only small quantities monthly.

De Beers did not want to concur, but were finally persuaded. The Premier Mine, however, declined to fall in. It reverted to selling on its own. Furthermore its production now leapt upwards, overtaking De Beers'. But increased sales meant a fall in price. Four years earlier the average price was 23s per carat. By 1908 it had fallen to 14s 9d and by 1909 to 12s 6d. The sale price had thus been almost halved. De Beers regarded the Premier's policy as lunatic. In Kimberley they had always tried to keep prices up. Cullinan's crew near Pretoria seemed hell-bent on driving them into the ground. The profit margin had shrunk so acutely, it was generally believed in Kimberley that the Premier would ruin itself.

Not that the collapse of a business rival would worry a competitor if it could happen in isolation. But the dying fall of the Premier was bringing down diamond prices all round. The Premier's price-cutting was already pinching De Beers. If it continued it must seriously wound them. The prospect was frightening and the threat could only be removed entirely if De Beers bought out the Premier Mine. But the Premier were not yet sellers. Even if they had been, De Beers (whose top management was strangely optimistic about future prospects) had no great urge to buy.

The purchase was done for them by none other than Barnato Brothers under the leadership of Barney Barnato's brilliant nephew S. B. Joel, another South African millionaire who founded a horse-racing dynasty in Britain. Barnato's were still the largest shareholders both in De Beers and in Jagersfontein when in 1911 they secured control of the Premier Mine. At De Beers Annual General Meeting in 1916, Solly Joel could boast that he had brought his new colleagues on the board of the Premier 'into line with us, to know and to say that there is only one thing, and that one thing is reduced production and higher prices'.

It was the cry of the capitalist to which the rising tide of social revolution was going so seriously to object over the next fraught decade.

The serious South African diamond crisis temporarily subsided. A few weaker members of the Syndicate, whose nerves and liquidity were insufficiently strong, dropped out, declaring they no longer saw a sure future in diamonds. The Premier Mine had been bolstered by the Transvaal Government who had waived £250,000 owed to it by the Mine. But in Kimberley De Beers had to close the Dutoitspan mine to keep production down. Confidence had been gravely shaken and diamond economics were growing very complicated.

CHAPTER ◆ FIFTEEN

The German Diamond Fields

In 1908 the diamond industry received a lancing blow from an unexpected and unprotected quarter. In April, diamonds were suddenly discovered in German South West Africa, far to the west of Kimberley on the roaring Atlantic coast. As the first whispers of the new finds drifted in, the Prime Minister of the Cape gloomily announced: 'A discovery of a new diamond field would be a hideous calamity for us all.' The chairman of De Beers took the news more sanguinely, but that was because his natural optimism made him doubt if the Germans had unearthed a worthwhile field at all.

The first stone (a mere ¼ carat) had been casually picked up by an African labourer originally from Cape Colony who was working for a German railway official, Oberbahnmeister Stauch, on the line linking Lüderitzbucht and Keetmanshoop. Stauch brought in two other officials to go prospecting with him and when they applied vainly for financial backing to a Berlin bank, the news leaked out.

De Beers had falsely discounted the discoveries. The Germans had found a glittering treasure trove on the surface. A very rich alluvial field covered a wide area. There was another 'rush' across Africa and from Europe to stake claims and to form syndicates, and in Berlin 'The Diamond Régie of South West Africa' was formed in 1909 as a monopoly to market all production. But, unlike the composition of the London Syndicate, the German diamond producers in South West Africa were not represented on their Régie at all.

De Beers might try to shrug off their new rivals by depreciating the smallness of the South West African stones compared with their own, but the threat to the diamond market loomed overall.

It was true that the Germans adhered from the start to the principle of single-channel selling through the Régie, to which De Beers subscribed through the Syndicate. But the Régie was totally independent of the South African producers who thus had no control whatsoever over its prices. In the interests of economic trading some link ought to be effected between the two neighbouring but somewhat unfriendly countries. The rancour incurred by the German Kaiser's support of the two Boer republics in the South African War had not been forgotten. South West Africa and the Union were limbs of two competing Empires based on Berlin and London. Though the Kaiser and King George v were both grandsons of Queen Victoria, their military men were already planning war.

De Beers could either try to buy prospecting stakes in South West Africa, where one British firm was already ensconced, or try to buy shares in the German Colonial Company which controlled much of the country. Alternatively De Beers might buy South West diamonds directly as merchants. There was an optimum solution: finally, De Beers could try to persuade the Régie to link up with the Syndicate. The pursuit of any of these courses would require foresight, careful planning, considerable finance and diplomatic tact of the first degree.

The Régie was selling its stones to a syndicate in Antwerp which created a rival market to London's. Then in 1912, as the war clouds began to pile up over Europe, the German producers finally obtained representation on the Régie and the stones were put out to tender for sale to the highest bidder. By 1913 prices had risen by 50%. The German Government then grasped the necessity of limiting output to maintain prices: it set a production ceiling of 1 million carats for 1913. At that juncture the Syndicate slapped down a tender for half a million carats of the Régie's diamonds.

De Beers' backs were now against the wall: here was their only marketing outlet buying from the competing producers next door. It seemed that De Beers must join with the Syndicate to purchase South West African stones. It would be a particularly sensible move if the Germans would henceforward limit production still further.

The Union Government, itself anxious about the future, urged De Beers to confer with their German rivals. Early in 1914, as Europe's Great Powers started to flex their arms for war, De Beers suggested to the South African Government that they should inspect the production in German South West Africa.

Ernest Oppenheimer accompanied the investigating team to give his expert advice. As Dunkelsbuhlers' representative he had been in close touch with the Premier Mine during the 1907 crisis, he had regularly attended the De Beers and Jagersfontein Board meetings, and he had become a leading local personality before he was 30. Without flamboyance or obvious thrustfulness, this small, neat, dapper man, looking far more a quiet businessman than a diamond prospector, had established a high reputation for acumen and trustworthiness. Instinct had nudged him into local politics: as a result he became better known. He could state the diamond case in Kimberley. Then, on the next level, he could state Kimberley's case in the Cape Parliament.

Oppenheimer became a Councillor for Kimberley in 1908, the year the De Beers mine closed in the depression, and then Mayor of the combined Corporation of Beaconsfield and Kimberley for three years from 1912. The experience of local politics, of dealing with personalities and interests outside the tight pale of the diamond industry was invaluable for a young man who till then had spent his entire working life in a diamond merchant's office. It not only practised him in public speaking and committee work (he was a natural persuader) and paved the way to his seat in the Union Parliament; politics also widened his horizons and thus his contacts. His published paper, *Diamond Cutting*, had been well received. This had replied to Press demands for an export duty on rough diamonds to encourage a South African diamond-cutting industry. Now followed his report on the South West African diamond industry. This too was acclaimed at the time and its prescience was applauded later. His experience in South West Africa, in addition, proved personally beneficial: within five years he was deeply involved in the formation of the Consolidated Diamond Mines of South West Africa.

Just before Britain declared war on Germany on 4th August 1914, De Beers and the other South African producers had at last concluded a deal with the Régie. A pool of South African and South West African producers would sell to the Syndicate. The Syndicate would receive a sales commission plus a further commission on profits. 'Outside' diamonds up to a value of £500,000 could be bought by the Syndicate. Of the total bought by the Syndicate, De Beers' quota was to fall slightly to 48%, while the Régie's quota was to rise slightly to 22%. Arrangements had hardly been made when the outbreak of war, which included operations by South African forces against German forces in South West Africa, froze this international agreement.

On 4th August 1914, the Kimberley mine ceased working and all diamond mines closed for the duration of the war. Faced with little to do Oppenheimer immediately left Kimberley, went to the Rand and started on the mining of gold. Many existing South African businesses had interests in both diamonds and gold. Prospectors in Kimberley had moved to the Rand, then came the finance houses and administrations. War pushed another group of men eastwards, of which the most eminent was Ernest Oppenheimer. To the opportunity accidentally created he brought his brain and foresight with the capital backing of his firm. From these beginnings grew the Anglo American Corporation of South Africa.

The great combine developed from the original association between Dunkelsbuhlers and two gold-mining investment houses: Consolidated Mines Selection Company and the Rand Selection Company.* For the investor, gold-mining had two advantages over diamond-mining: there was a continuing stable market for the product; and its production and marketing was spread over a number of firms. There was no tight two-handed grip like that exercised by De Beers and the Syndicate over the mining and selling of diamonds. There was thus an urge to invest in gold-mining from large professional investment companies in the Union and in Britain.

Dunkelsbuhlers had run a Johannesburg business from the early days of the gold rush and in 1905 they merged with Consolidated Mines Selection, keeping control by retaining the appointment of the managing director and two other directors. Rand Selection, already predominantly owned by Consolidated Mines Selection, was drawn into the new combination.

Thus Ernest Oppenheimer of Dunkelsbuhlers, the former Mayor of Kimberley, came to be doing excellent work for Consolidated Mines Selection. The business often took him between London and South Africa, for British control in Consolidated Mines Selection had increased on the outbreak of war when its four German directors had summarily to retire. Oppenheimer now started to plan the formation of a new, larger company. It would use American money for two reasons: first, no finance would be readily forthcoming from war-locked Europe. Second, Oppenheimer had befriended the American mining expert W. L. Honnold, a close associate of U.S. President Herbert Hoover.

*Formerly known as the Coal Trust.

So far American investment in South Africa had been on a small, personal level. A number of American mining engineers had followed their prospectors into the new country and helped to open it up. But there had been no large-scale investment from the big American financial institutions. The need for further finance was apparent to Oppenheimer; his friendship with Honnold put that finance at his right hand. Knowing the right people at the right time was as vital to Oppenheimer's success as to anyone else's.

Although originally planned as an amalgamation of gold-mining houses, the Anglo American Corporation had, from its inception in May 1917, been ready to bring diamonds into its web as Ernest Oppenheimer (then aged 37) laid down in a letter to Honnold that month. He already envisaged something of the group's complex interlocking future. The Corporation was hailed by the world's press – 'American millions for the Rand' – when it was incorporated with an issued capital of £1 million.

Although Oppenheimer was deeply involved in Johannesburg during the war years, he kept in touch with Kimberley's dormant diamond industry, planning for its post-war future. For the first two years of the war De Beers ceased ordinary mining in Kimberley, then the administration of the industry stirred. Solly Joel sold out Barnato's controlling holding in the Premier Mine to De Beers in 1917. Thus De Beers finally acquired the grip they had always sought over their Transvaal rival. The South African Government then arranged that the marketing of all German South West diamonds should pass through the Syndicate on a profit-sharing basis until the war was over. Confidence bloomed in the industry over these new arrangements and lasted into the outbreak of peace. Nothing could brake the post-war boom, in spite of signs of menacing diamond production in the Belgian Congo and Portuguese Angola. Demand for diamonds was sparkling.

In 1919 a 5-year agreement was signed between the main Union of South African producers and the Administrator of South West Africa* who had been appointed by the Union Government to represent producers there. This controlled all production to quotas which had to be effected through the Syndicate. The rival Régie had been eliminated by war and the German defeat. But while these

*The former German colony had been placed by the terms of peace under South African control as a Mandated Territory.

negotiations were going on, Ernest Oppenheimer was moving faster behind the scenes. Through Anglo American and with the backing of the American bankers J. P. Morgan, he was quietly bringing about an amalgamation of principal producers in South West Africa.

The capital to finance Oppenheimer's new company, called the Consolidated Diamond Mines of South West Africa (shortened in the trade to Consoldia and then C.D.M.) came from Anglo American, Morgans and a syndicate of rich South Africans including Ernest Oppenheimer himself. Peace was less than a year young when Anglo American acquired, under the twitching noses of De Beers, this important new diamond-producing combine, elements of which had so threatened De Beers and other South African producers before the war. Oppenheimer and his supporters had seen that South West African diamond sales would again menace the South African industry, if they were not all brought under one wing.

This was the pattern of Oppenheimer's takeovers: to move quietly but from strength, and to increase his bargaining power with each move. He preferred to add units together retaining his interests in each, rather than use one company to gobble up others. He had put his first-hand knowledge of South West Africa to full effect, while De Beers were occupied at home. They had delayed too long over their own moves. They had believed wrongly that the Allies would not permit deals to be done with German owners in South West Africa.

It suited the British and South African Governments very well to have the complexity of German producers united under one hand with whom they could deal. They gave Oppenheimer their blessing. De Beers were beaten before they even started. And, fortunately for Oppenheimer, they did not want even a minority interest in Consolidated Diamond Mines. Oppenheimer's victory for his new Anglo American Corporation was complete.

CHAPTER ◆ SIXTEEN

The Union Diamond Board

Reflecting on his first large takeover, Oppenheimer said that he had been considerably impressed by the Barnato Bros. purchase into Premier and by De Beers' subsequent purchase of that holding from Solly Joel which gave them control. He admired not only Joel's acumen but his method: the private purchase of a large stake in a company which would turn out to be increasingly attractive to a rival. Oppenheimer's moves in the construction of Consolidated Diamond Mines pursued the pattern of the Joel-Premier-De Beers deals.

The original suggestion about South West Africa came from a Cape Town colleague of Oppenheimer's on Anglo American's board: the Hon. H. C. Hull. Hull also clinched the final deal on his mission to South West Africa, but it was Ernest Oppenheimer and his brother Louis who worked out the details of the complicated amalgamation and resale. Even then he was thinking years ahead. In 1920 he obtained CDM's assurance that should their arrangements with the London Diamond Syndicate come to an end, then the Oppenheimer brothers had the first chance themselves to sell the company's output. At the time the eventuality seemed remote to the general observer. To Oppenheimer it seemed a probability for which he must carefully plan. It was going to prove, without cost to him, a powerful card to play in the future.

King George v's New Year's Honours List for 1921 contained a Knighthood for Ernest Oppenheimer. Primarily in recognition of his leading role in recruiting during the war, it set the Establishment's seal on the 'Kimberley Man', as he called himself when he stood as Member for that city in 1924. He held the seat for General Smuts' South African Party for fourteen years until 1938 and the eve of another European war. His political position was of great value to his business colleagues, for the Union Government since 1918 had intruded more and more into the affairs of the diamond industry. Oppenheimer was at hand to speak up boldly and plainly for the industry's interests.

The Government introduced a Bill to create a Union Diamond Board which would 'drive a wedge into the Syndicate'. This Board was to have powers to 'demand and receive diamonds from any producers named by the Board ... and to create a monopoly of sale and export through the Board'. The tone of the Bill and the dictatorial powers it proposed to invest in the Board thoroughly alarmed South Africa's diamond men.

But the Bill's conditions specifically excluded alluvial diamonds, an omission which was to prove important. After the early days and the development of the expensive deep mines, there were no big alluvial workings in the Union. There were plenty of riverbanks around where the odd diamond might be washed up, but they were uneconomic for the large mining companies to work. They were left to handfuls of tough, rough-living private prospectors.

Alluvial diggers in the Union were therefore generally little men eking out a precarious existence on their own outside the orbit of the great businesses. All parties were agreed that these little men needed protection; but none of the diamond producers supported the proposed Control Board. They resented the implied threat of confiscation and an Opposition speaker sarcastically enquired if the Minister of Mines had learned this trick in Moscow.

General Botha, a foe of Britain's in the South African War, was now Prime Minister of the British Dominion, so quickly had old scores been forgotten by big men. He was determined to build up a new domestic diamond-cutting industry in South Africa. Most of the world's cutting still took place in Amsterdam and Antwerp, but there were smaller centres in Germany, New York, London and Paris. It seemed sense to encourage a domestic industry by levying a duty on all stones exported uncut. The theory was sound: to avoid the duty Union producers would get their stones cut by South Africa's new cutters. In practice, however, the producers reacted violently. They were implacably opposed to any duty increasing the price of their exports. They pointed out through Ernest Oppenheimer the great difficulty in valuing uncut stones for duty. And they were incensed when they discovered the Minister of Mines proposed to give himself powers to 'require' producers to supply the cutters with diamonds, on terms he would fix, even if their sale elsewhere had already been agreed.

The Bill was complicated in structure and muddled in thought, and Ernest Oppenheimer methodically tore it to shreds in speeches in June 1926 and February 1927.

Oppenheimer was doing more than act as parliamentary spokesman for the diamond industry. He was aspiring to its unification. He could not yet challenge De Beers with Anglo American. In the Syndicate Dunkelsbuhlers could not yet speak with the same voice as Breitmeyers (who had succeeded Wernher, Beit) or Barnato Bros. It seemed to Oppenheimer (whose letters to his brother Louis harped on the theme) that an alliance with Solly Joel of Barnato's was going to be vital for the next leap forward.

As it happened, the desired alliance was brought about from the outside. Pursuing the principle which had led him into South West Africa – that of buying into the opposition – Oppenheimer started to acquire stakes in diamond companies in Angola, the Congo and in West Africa. In nearly all these enterprises he worked through Anglo American and Dunkelsbuhlers in close association with Barnato's. At the start of 1929 Anglo American had an 8% interest in the Syndicate and had arranged to market Angolese diamonds. The following year it included West African diamonds. Anglo American were thus extending their activities to diamond merchanting, and Oppenheimer found much wrong with the Syndicate's way of selling diamonds. He continued to attack its policy of reducing prices and resented the unpleasantly non-compromising attitude of his associates in general and of Breitmeyers in particular.

Oppenheimer was young and brilliant, successful and growing in power: no such man finds all friends among older rivals. Breitmeyers were not the only people who had no desire to further Oppenheimer's progress. He recognized that this opposition within the Syndicate would lead to an impasse. He began to contemplate founding a new Syndicate with Barnato's help. But would Solly Joel and Barnato's come in with him?

The year 1924 was a singularly inharmonious one for the Syndicate. Its members quarrelled among themselves and to varying degrees with the Government (which wanted to meddle still more deeply) and with the producers (who always wanted higher prices). Oppenheimer had a dual position representing both the Syndicate and the producers. Speaking for CDM he plainly saw the producers' point of view. Production of 'outside' diamonds was growing. Their producers joined South Africans in clamouring for larger quotas.

The Administrator of South West Africa was particularly anxious to be able to expand the sales of CDM, but his demands were not accepted by the Syndicate. He fretted impatiently for a few months. Then, at the end of the year Oppenheimer got a message from his brother-in-law Leslie Pollak who had heard from the Government's technical adviser. The message reported that the Administrator, impatient of fruitless Syndicate conferences, would consider new proposals from Oppenheimer after 1st January 1925, when the existing contract would have expired. This was the contingency for which Oppenheimer had planned back in 1920. If the Syndicate's arrangements with CDM lapsed, Oppenheimer and his brother had the chance of selling CDM's output. Now he would have to work very fast indeed to set up his financial arrangements in time. He was, furthermore, doubly inhibited: his two firms, Anglo American and Dunkelsbuhlers, were members of the Syndicate. Acting for *them*, he had to be sure it was in their interests if he led them on his breakaway.

The Government then exerted a final squeeze. On 6th January 1925, the Minister of Mines sent an angry circular telegram to all diamond producers in South Africa. It was an absolute ultimatum. The Government, fed up with the delays in finalizing agreements between producers and Syndicate, gave the former six days' grace – until 'not later than noon on the twelfth instant, otherwise Government will adopt such course as it may deem fit untrammelled by any negotiations hitherto'.

In anticipation of such a crisis and on receipt of his brother-in-law Pollak's tip-off, Ernest Oppenheimer sailed for London, to rally his allies.

CHAPTER ◆ SEVENTEEN

The third Great Man in Diamonds

The crisis burst in January 1925. Despite a plea from Ernest Oppenheimer to the South African Government, part of the South West African diamond production by-passed the Syndicate and was sold direct to an Antwerp merchant. The frame-work was threatened. After consulting his brother Louis and his bankers, Morgans, Oppenheimer cabled the South African Government. He offered to buy the output of the Consolidated Diamond Mines of South West Africa for Anglo American. The Government accepted the bid on 19th January. Almost immediately the Syndicate riposted. It expelled Anglo American 'for assisting South West African producers and thus preventing the Syndicate from doing a favourable deal with Union' (i.e. South African) 'producers'. Dunkelsbuhlers were also expelled for having supported Anglo American's action. It was a public declaration of war.

Unknown to the Syndicate Oppenheimer had tried to bring off an even greater coup by carrying his expansion into his rival's country. He had secretly also bid the Government for the complete production of all the Union's mines. Lest he be accused of acting behind the backs of his colleagues, he offered them participation in any deal he might be able to fix. But to secure a fair price he dared not reveal before the deal the size of his bid for the South African production. Oppenheimer's bid price was however leaked to the Syndicate, which raised its price and obtained the output of all the major South African producers including De Beers. So there were now two selling channels in action: Oppenheimer's and the old Syndicate's. It was the position Oppenheimer had for so long sought to avoid. However he regarded the situation purely tactically. It had to form only an incident in his strategic aims, because his contract with CDM and the Syndicate's contract with De Beers were both only for the short term.

During the year Oppenheimer moved like a general to establish his strategic position. First, he negotiated a contract with the Administrator of South West Africa, substituting a new 5-year contract for his short-term one. His syndicate to finance these purchases comprised Dunkelsbuhlers, Anglo American and his two banking allies: Pierpont Morgan in New York and Morgan Grenfell in the City of London. But he was also in partnership with Solly Joel's Barnato Brothers for diamond buying from the Congo, Angola and West Africa. And Barnato's were still members of the old Syndicate. Here was the link between the rivals.

The link, what was more, might well be turned to Oppenheimer's advantage. Solly Joel, a director of De Beers and representing the largest shareholders, had for some time been dissatisfied with the chairman. Joel, like Oppenheimer, believed that diamond selling should revert to a single channel. But should that be the old Syndicate? Or Oppenheimer's new one? Solly Joel's intentions were vitally important. And still in doubt.

A message reached the Oppenheimer brothers suggesting they rejoined the old Syndicate on a new 5-year contract. But Ernest and Louis were not to be tempted. They replied that without satisfactory new proposals from the old Syndicate – ('after all, they turned us out') – the Oppenheimers would make both the Government and the diamond producers an offer for a 5-year contract. They would do this even if Solly Joel did not bring Barnato's in with them.

Walter Dunkels of Dunkelsbuhlers in London supported Oppenheimer. He reckoned that if he stood firmly aloof from the old Syndicate Joel would weaken and cross over. By the middle of the year the Oppenheimers were ready to put their terms to Barnato's. Solly Joel quickly agreed in principle, leaving precise terms in abeyance. Now they had to move swiftly to forestall the Syndicate's bid for De Beers' production. On 16th July Oppenheimer, on behalf of Anglo American, Barnato's and some other friends, made his offer to De Beers for the entire production of De Beers and of the Premier Mine. De Beers argued for fourteen days and then accepted.

With these in his hands Oppenheimer was nearly there, his allies congratulated him on his coup and in October the old Syndicate finally went out of business and sold out to him. Everything went smoothly. The outlay to buy the old Syndicate's stocks of diamonds was not so heavy as had been feared, because sales were going briskly. Simultaneously Oppenheimer held a producers' conference which unanimously accepted the inter-producer's agreement and the agreement for sales to his Syndicate. The new Syndicate set up its new offices in Kimberley on New Year's Day 1926, comprising Barnato's with 45%, Anglo American and Dunkelsbuhlers sharing 45%, and the Johannesburg Consolidated Investment Company with 10%.

Oppenheimer next accomplished something on which he had set his heart back in the war years. He had wanted for 10 years to join the Board of De Beers, but his overtures had always been politely rejected. Finally in July 1926 he was elected a director. He was halfway to his ultimate goal. His basic theme was still the unification of all the means of producing diamonds, then of selling diamonds in the Union of South Africa and in South West Africa. He had it in mind, too, to bring into the selling monopoly diamonds produced by outside sources in the Congo, Angola, Ghana and even further afield, if that could be contrived. De Beers seemed to Oppenheimer to be the only vehicle which could carry such a combine. But to control De Beers he must be at the steering wheel. What he now wanted was the chairmanship. For the next three years he struggled for the top seat.

But the diamond world was now subjected to one of its frequent convulsions. Two new areas of alluvial diamonds, rich and large, were suddenly discovered far apart: one in the Lichtenburg district of the Transvaal; the other on the bleak and hostile shores of the Atlantic in what was called Namaqualand, south of the Orange River which was the frontier with South West Africa.

The double discovery in 1926 and 1927 hit the diamond industry like the twin blows of torpedoes from port and starboard. It shuddered. This time no general world depression had caused the crash in the diamond market. The crisis was of its own making: uncontrolled over-production, the danger against which Rhodes had preached in the old days, against which Oppenheimer was now constantly inveighing. Now the blunder of excluding *alluvial* diamonds, despite Oppenheimer's warnings, from the Government's 1925 Diamond Control Act came boomeranging back. The Government found it had no control at all over these sparkling new alluvial discoveries. It had excluded itself deliberately, and was now impotent. The spate of poor diggers pouring down on the Lichtenburg farms and Namaqualand shores, had not the finance to hold onto stones and wait for stronger markets. What they found they had to sell immediately. An 'Aladdin's Cave' of diamonds, as Doctor Merensky called his great hoard at Alexander Bay, began to flood the market. The prices slipped, and started to plunge desperately.

Lichtenburg was so close to the Premier Mine that the finds on its farms hit the mine particularly badly. Kimberley in its turn suffered particularly from the port torpedo: the alluvial stones coming in from the Atlantic coast were of dazzlingly high quality; they were quite a match of their own. The CDM however, though equally hit by the deluge of stones on the market, had strong grounds for hoping that the Namaqualand fields might well extend north of the Orange River into South West Africa and thus lie in their own territory.

The new discoveries in South Africa reverberated in Europe and America. Diamonds now were coming in up to 30% cheaper. If two fields could be discovered in two years as far apart why, reasoned the distant dealers, should not a host of new fields be on the verge of exploitation? Surely the right thing to do was to hold off buying. The prices plunged on downwards.

To try to bolster confidence in the trade the Syndicate was already having to buy £40,000 worth of Lichtenburg diamonds each week. They were appalled to learn that future Lichtenburg production was forecast at no less than £100,000 per week. At that rate the Syndicate's reserves would soon be exhausted as the stones piled up, losing their value daily.

With no control over the new production, the industry attempted to rationalize the marketing of diamonds. Following the lead of Dunkelsbuhlers in London a combine was formed of all the direct importers of South African alluvial diamonds in London and Antwerp. This combine undertook to sell at least part and if possible all their collected shipments through the London Diamond Syndicate. A measure of control was thus effected, but the London Syndicate had to keep on buying alluvial stones to hold.

During the two crisis years of 1926 and 1927, Oppenheimer was striving for the chairmanship of De Beers, speaking in Parliament and talking to the Minister of Mines about the crying need for Government control over alluvial diamond production and, in co-operation with Solly Joel, taking a stake himself in the Lichtenburg area by buying the most potentially valuable farms there.

This was Oppenheimer's classic expansionist move repeated: to move personally, and outside his main business interests, into new competitive areas. He had done this already in other parts of Africa. Now he moved again closer home. He and Solly Joel knew the value of the Lichtenburg area must rise and that at some time it must be brought under the wing of a large combine, which had to be De Beers. The Joel-Oppenheimer stake in Lichtenburg therefore, served three functions: it maintained an interest in the opposition; it was an appreciating capital investment; and it afforded a future bargaining card for a merger.

To the west in Namaqualand large numbers of the first discoverers had now sold out to Dr. Merensky's group, called the 'H. M. Association', in which Sir Abe Bailey was involved. The Diamond Syndicate's members, Dunkelsbuhlers and Barnato's, had also bought interests in Namaqualand and so had De Beers and Anglo American. With his interests in all four, Oppenheimer was in a splendid position to bring them together, and by the end of 1927 he had accomplished this merger to form the Cape Coast Exploration Company.

But Oppenheimer wished to expand further. He had difficulty in convincing his brother and his other backers of the importance of the Namaqualand discoveries. But in the end, after long negotiations he overcame Solly Joel's reluctance to plunge in with him, and the Oppenheimer-Barnato combine bought out all the Merensky shares for £1,103,750. The deal had been complicated by Dr. Merensky's understandable insistence that it should be for a round million pounds plus the £103,750 already advanced to him under a pooling agreement.

In industrial politics Sir Ernest Oppenheimer's position was now immensely strong. But his personal aims were still frustrated. Although he was now recognized as *the* man in diamonds there was a barbed reluctance to accept the Syndicate leader as chairman of De Beers. There was a natural doubt whether such a step, forming a personal link between a vertical monopoly of producing and selling, would be in the interests of the industry. For the same reason Oppenheimer could not persuade the host of alluvial diamond producers to combine under De Beers as one body. It looked, as Lord Bessborough of Morgan Grenfells suggested, as if there might be a compromise: one combine for mining, and another for alluvial. But here the bounds of definition were confused. In the rivers 'pipes' were being discovered which had seemed pot holes. These pipes required underground mining by large companies. And the diggers were not all still independent poor men. Many had banded themselves into companies with outside shareholders and considerable capital. The original purpose in excluding alluvial diamonds from state control to give some freedom to 'the little man' had largely been washed away.

At Kimberley the struggle of power politics continued. De Beers remained the greatest company with the famous name, but the personal leadership of the diamond world had indisputably passed to Sir Ernest Oppenheimer. In order to regain full command of the situation De Beers needed to negotiate a new relationship with Oppenheimer. But these negotiations meant bringing up the vexed question of the chairmanship. Oppenheimer insisted that his securing the chair was a *sine qua non* of any arrangement. The Board of De Beers were reluctant to appoint him. It was uncertain how far either De Beers' London bankers or Rothschilds acting for their French shareholders really supported his candidature. Nor on this point was Oppenheimer sure even of Solly Joel's backing. Sometimes Joel seemed strongly behind him, at others he refused to move against the existing chairman, his old friend that veteran Kimberley figure, Sir David Harris. Oppenheimer felt that Harris had never liked him since the day when Oppenheimer had failed to get a seat on the board of Jagersfontein. That rejection still rankled. He could not bear the thought, as was suggested by Solly Joel, that David Harris should himself decide whether or not to make way for him. This would once again put his future and his pride in Harris' hands. In such circumstances he would prefer to stand down altogether.

Discussions towards a new Diamond Board of twelve members, half from De Beers including the chairman, and half from the Syndicate collapsed on the old grounds: the double interests of Oppenheimer and Joel with the selling side of the industry. This time Oppenheimer decided to take this reaction as a personal affront to his and Solly Joel's honour. Words like 'mistrust', 'direct insult' were cabled to and fro. The situation had not been resolved by 1929, and Oppenheimer threatened privately to resign from De Beers.

Morgan Grenfell now intervened, proposing, in March 1929, a new buying and selling company with twelve directors. Of these

De Beers would appoint three; Joel and Oppenheimer together six; the Premier Mine one, and the two bankers (Morgan Grenfells and Rothschilds) two more who would be public men outside the diamond industry. The bankers were open to suggestions from Joel and Oppenheimer about the chairmanship ...

The situation looked promising for a deal and outside economic pressures combined to remind the participants of the importance of unity. There was a considerable falling off in the demand for diamonds, world production had mounted and the Syndicate was holding gigantic stocks of Lichtenburg diamonds and of the Merensky hoard. Everything suggested the hour was at hand when a national control over South Africa's diamond production and sales must be instituted. But the Namaqualand diamond stock-pile which encouraged the Syndicate to do a deal, simultaneously turned De Beers against one. The Namaqualand high-quality stones would be in direct competition with their own.

Oppenheimer made his counter-proposals to those of Morgan Grenfell's. He suggested enlarging the Syndicate into a Diamond Corporation embracing producers and the existing Syndicate equally. To give De Beers final control it was agreed that they would have the right to appoint the chairman of the Diamond Corporation, and also the right to remove him at any time, and to appoint another director in his place. This was a strong card and an attractive inducement to De Beers.

The Diamond Corporation would deal with the acquisition of all 'outside' diamonds which had formerly been handled both by the Syndicate and also by producers like De Beers. Syndicate and producers would now surrender these rights to the new Corporation.

At last all parties saw an arrangement in which all benefited in part and the South African diamond industry as a whole. Years of discussion and the exercise of power were at last coming to fruition. The seeds had been planted in Kimberley's early days when Rhodes and Barnato battled for power in that shanty town.

The basis was agreed. The Diamond Corporation Ltd. was incorporated on 18th February 1930. Its existence proclaimed the open partnership between producers and sellers and so removed at last the old objection to Sir Ernest Oppenheimer being chairman of De Beers. On 20th December 1929, he was elected to the chairmanship, and in March 1930, he added the second barrel when he became chairman of the Diamond Corporation. He was thus King Emperor of the diamond world.

But as he came to his twin thrones the empire started to crack, for in October 1929 the New York Stock Exchange had begun to crash. The world depression of the Thirties loomed.

Oppenheimer was fifty in 1930 and the next five years were the toughest of his life. They saw the collapse of world economies. Sales of diamonds crashed. The industry had to build up larger and larger stocks to top those already accumulated from Lichtenburg and Namaqualand. But potential buyers, knowing of these vast stocks, had further cause to hold off buying. The only solution was first to restrict, then to cease production altogether. In 1932 all the mines of the De Beers group had closed. Not till 1935 did some mines resume operations on a limited scale. There was heavy unemployment, economic distress and bitter battles with the Government. It was not until after the end of the Second World War that the Diamond Corporation finally disposed of its gigantic stocks.

Fortunately the foundation of the diamond industry of South Africa had been soundly accomplished by Ernest Oppenheimer in 1930. As a result it could survive depressions and war, the emergence of Russian diamonds, and his death. The astonishing growth and all the organization of the industry sprang from the base formed by Barnato and Rhodes, then by Oppenheimer, the three big names in the diamond world of South Africa.

CHAPTER ◆ EIGHTEEN

A Sellers' Monopoly and Artificial Diamonds

Cecil Rhodes' conviction that those early forces of production at Kimberley must consolidate, has grown into an organization which controls the production of diamonds in all southern Africa. It is now joined with a selling organization which practically controls the selling of the diamonds of the world. 'Diamond production is a good thing' said Cecil Rhodes, 'if it is rightly handled, but if there is over-production it brings misery and disaster to all.' Thirty years after his death, Sir Ernest Oppenheimer, another chairman of De Beers, concluded the rationalization of the industry.

The Diamond Corporation which he founded is now a subsidiary of De Beers. It is still engaged in the purchase of diamonds (or 'goods', as the trade has it) for outside producers, but the marketing of their stones was taken over in 1934 by a new organization, the Diamond Trading Company. The Diamond Corporation, relating ıts purchases to its sales, still guarantees minimum purchases to producers so that the mines have a supported market even during times of depression.

An additional body emerged at the same time: the Diamond Producers Association of which the Diamond Corporation, the South African Government and De Beers are members. After the Second World War came a further company: Industrial Distributors Ltd., formed to market industrial diamonds as opposed to gem stones, and this company plus the Diamond Corporation and the Diamond Trading Company constitute the kernel of the Central Selling Organization.

Such an organization must of its essence be a target for opponents of monopolies. From time to time accusing fingure have been pointed at De Beers and Anglo American who possess the means both to produde and to market the majority of the world's most desirable gem stones and its hardest mineral. In their defence De Beers have this to say officially:

'The policy of the Central Selling Organization is to ensure the utmost degree of stability – not an artificial high price level – in the price of diamonds.' They recognize that in most commodities competition is helpful, 'But the case of a luxury such as gem diamonds is different. They may be desirable but they are not essential and if their price were to be subject to violent fluctuations, people might well decide to do without them.'

They remind us that diamonds are also desired for investment purposes. 'They never wear out and are, therefore, looked upon as a permanent asset. The annual production of diamonds represents a comparatively small proportion of the total quantity of diamonds held by the public and it is necessary not only to obtain stable prices for the current output of the mines, but also to protect the value of the huge quantities of diamonds which are owned in the form of jewellery throughout the world. No commodity exists in which the value of the raw material commands such a high proportion of the value of the final product as in the case of diamond jewellery. Fluctuations in the price of rough diamonds would, therefore, be directly reflected in the market value of jewellery.'

So much for the protection of the public. Now as to means. 'To maintain price stability, it is necessary in times of depression to stock quantities of diamonds. Such a policy clearly requires large cash reserves and the companies within the Central Selling Organization have consequently set aside substantial cash reserves for the protection of the trade. In times of prosperity the situation is, of course, very different. The demand for diamonds may be such as to allow producers outside the Central Selling Organization to procure higher prices for their diamonds; they will merely seek to obtain the advantages without paying the price for that degree of stability which the Central Selling Organization brings about. If there were too many producers acting in this manner it would lead to the collapse of the entire marketing system and a crisis in the trade. It is, therefore, essential that the Central Selling Organization should handle the greater part of the world production.'

De Beers are sensitive to continuing criticism. 'The policies adopted by the Central Selling Organization have given rise to several myths and misconceptions. It is often stated' (which indeed it is) 'that the producing companies deliberately mine in such a way as to restrict production to a level calculated to cause an artificial shortage and keep the price of diamonds at an unreasonably high level. This can be refuted, firstly, by recalling that the contracts between the Diamond Corporation and the producers try to ensure continuity of mining operations in bad times as well as good. Secondly, the mines are at the present time not only working at their maximum capacity but in many cases have expended huge sums in modernizing and expanding their plants.'

Their case rests. It is as good a defence as any for the maintenance of a monopoly, even if, as some critics will think, they do protest too much. But then, nobody *has* to buy a diamond, has he? He may even make one himself.

In 1955 the General Electric Company of America successfully synthesized diamond, and at first caused dismay in the natural diamond industry. So far, synthetic diamond grit remains no more than grit and can be used only industrially for grinding.

Mr. Harry Oppenheimer believes there is room for both natural and artificial diamond. De Beers scientists followed the Americans in synthesizing diamond and further work in this field is being carried out under maximum security conditions in several buildings at the Crown Mines near Johannesburg.

Their Research Manager there, Dr. F. A. Raal, tried to explain synthetic diamonds to me:

'To convert graphite or carbon to diamond you need a high temperature coincidental with a high pressure. During the nineteenth century, many attempts were made to synthesize diamond like this. Then in 1904 Henry Moissan, a Paris Professor, announced that he had succeeded. His method was to heat a crucible containing carbon and iron in an electric furnace to a temperature of about 3,000°C and then to quench this rapidly in water. He argued that, if a metal was chosen which would expand on solidification, then very high pressures would be produced in the centre of the solidifying mass because of expansion of the metal against the already solid outer skin.

'Much of the criticism levelled against Moissan's claim is based on the argument that he could not possibly have achieved the temperature and pressure needed for thermo-dynamic stability of diamond. But Moissan's results, taken at their face value, strongly suggest that he did make diamonds. It is unfortunate that none of his material has survived, because modern techniques, such as X-ray diffraction, would have made identification positive.'

James Ballantyne Hannay's claim to have synthesized diamond is unique; for some of his material, twelve crystals in all, has survived and is in the British Museum. Hannay, a Glasgow chemist and fellow of the Royal Society of Edinburgh, conducted a series of experiments in 1878 based on the fact that, when a gas consisting essentially of carbon and hydrogen is heated under pressure in the presence of a metal such as sodium or lithium, the hydrogen would combine with the metal.

He reasoned that the carbon left behind might then well crystallize out as diamond. To obtain the high pressure and high temperature Hannay used tubes of iron sealed by welding. They were then heated in a large furnace. Many explosions occurred. Finally he claimed success after using a mixture of rectified bone oil, paraffin and lithium. In the hard black mass lining the crucible he found small transparent diamond-like crystals, which, after careful examination by the Keeper of Minerals in the British Museum were pronounced diamonds.

In 1943 these crystals were examined with the aid of X-rays and it was shown that eleven of the twelve crystals were indeed diamonds. Speculation is still rife: did Hannay really make those diamonds? Or were his mixtures secretly 'salted' by an over-zealous assistant?

The first authenticated synthetic diamonds were made in February, 1953, in Sweden* after twenty-three years of design and development work. Some forty small crystals were produced at an estimated cost of £150,000. De Beers followed the General Electric Company of America in 1958. The Swedish company failed to file patents for its process and the General Electric Company could not do so for a time by virtue of a USA Government secrecy act.

Synthetic diamonds are now also manufactured in Germany, Czechoslovakia, Russia and Japan. De Beers has established large synthetic diamond plants in Springs, South Africa and also in Shannon, Eire, where diamond is manufactured and processed today for use in nearly every industrialized country in the world.'

Dr. Raal explained how to make a synthetic diamond. High temperature is achieved by electrical means. High pressure is not too

*By Allmänna Svenska Elektriska Aktiebolaget.

difficult: the obvious thing to do is to apply a large load over a relatively small area. To obtain high pressure within the bore of the die where diamond is made, an inner capsule is machined from pyrophyllite, commonly known as 'wonderstone', which is mined in South Africa, and will not easily flow under pressure. Apart from its unique pressure characteristics, wonderstone also serves the important purpose of electrically insulating the anvils from the die. This is essential because a large current is sent through the capsule to obtain the high temperature required for diamond synthesis.

First the wonderstone capsule is shaped to a set design. The capsule is then loaded with graphite and a metal – such as iron or nickel. The capsule, with its contents, is then inserted into the bore of the die. Dr. Raal continued, 'When load is applied in a hydraulic press, the anvils move in, the wonderstone is extruded to form a pressure-tight seal and the graphite and metal combination is compressed to a pressure of about 60 kilibars. Then a large electrical current of about three thousand ampères is sent through the compressed mixture. The temperature builds up almost instantaneously to the required 1,500°C.'

The metal melts and dissolves some of the graphite to form a carbide, out of which diamond crystallizes. (The whole operation takes no more than five minutes.)

The compressed capsule is pushed out of the bore and subjected to light crushing. The metal and wonderstone is removed by hand picking and the diamond is recovered from the remainder by chemical treatment.

Dr. Raal concluded coolly, 'Synthetic diamond is only useful for industrial purposes. It's worthless as jewellery because it's so small and because of its greenish-yellow or greyish-black colour. Man hasn't yet completely succeeded in emulating nature. The largest synthetic diamond crystals grown are only about one millimetre in size!'

CHAPTER ◆ NINETEEN

A look around the new Finsch Mine

Little more than a century after that first diamond was being kicked around by the children of Mevrou Jacobs, the city of Kimberley glitters on the veld. Where the bare farmlands rolled, live nearly 100,000 people: 27,000 Europeans, 43,000 Bantu, 20,000 Coloured and 1,100 Asiatics. Yet brick buildings only began to replace the diggers' tents in 1873.

A neat airport which pioneered South African cross-country flights now lies south of the city. Jets whistle in on their route between Cape Town and Johannesburg, and a cluster of De Beers' private planes wait to wing managers, geologists and visiting English writers to the Finsch mine, or to the Premier, out to the fields in Botswana, back to the office in Johannesburg. The Boer War battlefield of Magersfontein is commemorated with a museum and in Dutoitspan Road, along with the splendid old Kimberley Club, stands an equestrian statue of Cecil Rhodes, the 1914–18 War Memorial and Kimberley House where the Diamond Trading Company stores its goods for sale to London (for world-wide distribution) and to South African diamond cutters.

Five miles out on Carter Ridge a monument marks the spot where Colonel Scott-Turner and twenty-one men were killed trying to destroy a Boer siege gun. Beneath the Honoured Dead Memorial on Memorial Hill (built of stone brought from the Matopos Hills near Rhodes' grave) lie other fallen soldiers.

Not only fighting is commemorated. The 'Big Hole', the Kimberley Open Mine based on that old Colesburg kopje, first unearthed by Fleetwood Rawstorne's drunken servant, gapes darkly on the city's edge. The greatest man-made hole on earth is silent now, but it worked for 44 years, yielded 21 million tons of 'blue ground' and produced no less than 3 tons of diamonds.

By its side you take tea in Cecil Rhodes' American-built Pullman coach, all mahogany, copper, brass, five sleeping bunks, and an ornate lavatory. You walk through reconstructions of Barney Barnato's Boxing Academy where that strange, bouncing little man kept fit, and take a turn in Sir David Harris' ballroom resurrected as it was when it became the coveted apex of Kimberley's social life.

But only a dart away in a De Beers' small plane is the new diamond world. Out near Postmasburg 130 miles from Kimberley is De Beers' new Finsch Mine, product of the latest great coup of individual prospectors who, after years of searching, finally struck it rich. It is the most important pipe uncovered in South Africa since the Premier's discovery in 1902.

Two prospectors, Allister Thornton Fincham and his friend Ernest Schwabel, were prospecting for asbestos in 1958 along the stony ridge which crossed the farm Brits. Fincham pegged four claims and Schwabel built a cottage. The cottage turned out to be right on the brink of a huge diamond 'pipe', and one of Fincham's claims covered its area.

The unearthing of garnets had first hinted faintly to the two prospectors of the possibility of diamonds. They had been searching most of their lives: they always hoped, but they knew how minute the chance was of a really great strike. But as they dug holes to find asbestos, Schwabel kept coming across garnets. As they went on, their discovery gradually presented the pattern of a large circle. They might just be delineating a colossal diamondiferous pipe. Then markedly denser vegetation over the area on the ridge suggested the presence of water retained by diamondiferous kimberlite. Fincham took a gamble, ignored his asbestos venture, applied for a permit to prospect for precious stones, formed Finsch Diamonds with Schwabel and blocked other applicants for prospecting rights. In November 1961 work began sinking pits. The first wash brought up a diamond of $\frac{3}{4}$ of a carat. Nine more pits were sunk: all yielded diamonds. In the first month of mining, $33\frac{1}{2}$ carats worth of diamonds were recovered. The excitement reverberated in waves from that remote bleak ridge all over the diamond world.

Only a year later the pipe's yield had shot up a hundred-fold: 50 carats every day were now coming up. De Beers, who had naturally kept in the closest touch with developments, persuaded Fincham to let them prospect the pipe down to a depth of 200 feet. Fincham and

Schwabel had neither the capital, heavy equipment nor time, to go down further than 40 feet. The result of the probe was staggering. In the month of May 1963, the pipe produced diamonds weighing 1,634 carats. De Beers bid. The two prospectors accepted. De Beers bought out the entire share capital of Finsch Diamonds for £2¼ million. Of this Fincham took 40% directly and a further 15% through a shareholders' company controlled by him. Figuratively overnight, and certainly within 18 months, Mr. Fincham had become another, and the latest, diamond millionaire.

But the land is what in South Africa is still called 'Unalienable Crown Land'. This means that De Beers, like Fincham before them, cannot buy it outright, but have to operate under a government lease. Its terms provide that, after repayment of the initial capital, De Beers pay the state 60% of all profits. Of the balance remaining for De Beers, two-fifths is payable to Finsch Diamonds in payment for the transfer of their Government lease. The mine is very big business now. It produces about 2 million carats of diamonds a year and employs 500 people out in what was empty veld a decade ago.

A mountain of earth hundreds of feet high looms against the azure sky. It has been gnawed by great machines out of the hole which, gaping like an orange wound in the bushy veld, is now half a mile wide. In its first seven years of full-scale excavation the hole has already sunk down 300 feet deep. Nearly 2 million tons of yellow ground have been excavated from it and hauled aside for washing and recovery. It seems that the blue ground almost certainly extends below the yellow to at least 1,000 feet. This would represent a further quarter of a century's open mining before work would have to go underground. At the moment 27,000 tons of rock are blasted and hauled out daily.

Sometimes the rain washes stones from the huge mound full of diamonds, and visitors or African workers pick them up. Since the old regulations against IDB are still just as firmly enforced no unauthorized person can legally keep a stone. But rewards are dished out liberally for finding stones outside the normal work area. A cheque for £123 was presented to a lady just before our visit for 'quite a nice stone' which she had picked up in the car park just as she was getting into her car. In the previous week there had been fourteen 'pick-ups' by African workers; the largest, a 12-carat stone, rewarding the finder with £300.

In addition to the £2 million spent on washing and recovery plant, a new town has been built for Finsch Mine on the edge of the bare escarpment. Inside the screened and guarded perimeter are modern equivalents of the old compound described on page 60 by the *Pall Mall Gazette.* In the Finsch compound 360 African workers live entirely for their period of service. This may be for three, seven or nine months at the most. Though many Africans want to work a longer contract they are not allowed to spend longer than nine months in the compound without a holiday. Conditions and pay attract labour. Every Monday there is a queue of about seventy Africans who have made their long way there from all parts of southern Africa inside and outside the Republic, hoping for jobs. There are seldom more than ten vacancies and preference is always given to former workers returning after their holiday.

The result is a work-force of picked Africans, for whom the lure of good bed and board and a relatively high wage corresponds to that of the Army for the unemployed in Britain in the nineteenth century. The Africans at Finsch are, for Africa, particularly skilled. They are operating machinery on their own and beyond the eye of European surveillance, a situation which rarely obtains in South Africa. Pay averages about £1.50 daily and the food sold in the compound's own shops is subsidized. Facilities for sport are provided; the doctor expects a 7lb to 10lb weight increase in each African worker during his stay.

Three 8-hour shifts are worked. The normal term of service is seven months in the compound, after which there is an intensive X-ray check on every man proceeding on his four months' leave, in case any diamonds have adhered to his clothes or person. The Africans' leave usually entails an enormous spending spree back with their distant families in their poor villages and townships. Then most of the workers come back again standing in the queue on a Monday morning. 'Usually they're penniless when they return', says the manager.

The compound resembles a spick and span Army barracks of 60 blocks without any obvious drill sergeants. Twelve workers sleep and live in each 'barrack room', in which only senior hands have their own rooms. Each building has its own bathrooms and eating places, but as the Africans generally prefer to cook for themselves, there are separate long cookhouses with rows of stoves.

The workers reflect South Africa's mixture of tribes and religions: the Xosas making up 40%, Orange Free State Basutos 11% and Zulus only 2%. Religions are Methodist 31%; Anglican 19%; London Mission Society 13%; Dutch Reformed Church 6%, and Roman Catholic 5%. A number of the Xosas adhere to their own tribal religions. All the Xosas and some of the rest speak Afrikaans as well as their own languages. The common tongue is a pidgin-Bantu called 'Fanakola', which was developed in the goldfields.

Next to the compound the recovery buildings start, huge white structures with sloping roofs. From their upper floors long necks, like giraffes grazing, stretch to the ground. Up these, the crude ore is escalated. The earth, blasted out of the levels which descend the hole in 40-foot giant steps, is fed by mechanical shovels and feeders into dump trucks. Their burden is tipped into a huge reception bin, crushed down to 1¼ inches and then passed onto the washing plant.

Here rotary pans, resembling the prospectors' original washing pans except they are larger and made of metal, add 'puddle' to the crushed ore. The muddy liquid is stirred around by crude-looking radial rakes whose teeth gyrate the mixture so that light materials float off outwards, and the diamonds and heavy stones sink gradually to the bottom. A series of washing pans completes this operation and the concentrates are moved to the Recovery Section. About 11,500 tons of material are washed daily. Only 300 tons, or just over 2%, move on to Final Recovery.

There, any remaining ironstone particles are separated magnetically and the residue is washed again and split into two sizes: above ¼ inch and below. Once again the lighter materials are floated off and the concentrates pass into long drums called 'attrition mills'. These are filled with little steel balls which, whirling round as the mills spin, scrub off the grit and grime still sticking to the diamonds.

The remnants are then plunged into an adhesive reagent which sticks onto most of the diamonds. The mixture then is shaken slowly onto grease-belts, exactly as in the original method. And on the grease-belts, vibrating horizontally as the water flows across them at right angles, the diamonds (and a few other pieces of residue) finally stick. Colour-sorting machines check over the products of the grease-belt to remove further waste, and the tailings which have passed over the grease-belts go finally through an optical separator to catch any diamonds which failed to stick. The reflection of a beam of light by a diamond going through the Gunson sorter automatically sends a signal to an air nozzle below the sorter. This blasts the falling diamond out of the falling trickle of tailings.

The diamonds from all the grease-belts and optical separators are then simply collected and sent to Kimberley for sorting. A quarter of the Finsch output is generally of gem quality. The rest makes industrial stones.

CHAPTER ◆ TWENTY

Diamond in Action, and for Girls

Three and a half million carats of industrial diamonds are now produced annually by South Africa and South West Africa. Their uses cover the world – and space. A $\frac{3}{4}$-carat diamond had to be used for machine-polishing the heat shield of the Gemini space capsule. The highest quality carbide-cutting tools had failed to do the job, when the McDonnell Aircraft Corporation turned to the world's toughest mineral. The tool had to operate at white heat – about 3,500°F – and without any coolant other than air because of the risk of contamination. Experts deemed this an impossible task. But it worked. The stone, produced by the Industrial Diamond Co., did its job in three hours, and can machine three or four complete heat shields in its life.

Diamond saws cut the ceramic fibre insulation between the capsule's shells. The seven miles of wire in Gemini linking the control systems are drawn through diamond dies to produce the vitally precise gauging. Rocket engines, all the glass for the windows and covers, the nose cap, the synthetic jewel bearings in the controls, and the electronic devices with minute transistors are cut, ground, drilled and polished with diamond.

Diamond cuts the heat shields on the Minuteman Missile, each tool completing at least 35 miles of machining before it needs to be resharpened. Against this performance, a carbide-tipped tool has a maximum of one mile. Diamond has been used extensively for the same purposes in the construction in Britain and France of the Concordes.

On the domestic front, South African diamonds are used to cut ceramic tiles, to drill holes using diamond abrasives, and to grind ceramics flat or in forms. Bandsaws, circular saws and reciprocating saws are all available with diamond-impregnated edges for tasks requiring the toughest-cutting, longest-lasting mineral. Though most industrial diamonds are used for cutting in different ways, they also form machine components: styli for record-players, hardness-indenters, distance-stops and bearings. They are used for lenses and prisms and, in electronics, they provide insulating bearings and semi-conductors.

Diamond runs through the heat, din and dust of industry, just as it sparkles in the ring on the girl's finger – 'a Diamond is Forever' – in modern glittering earrings, and in the tiaras of the old nobility.

In building and construction, diamonds saw, drill, grind and polish all stones, pierce the hardest strata, saw bricks and concrete panels, drill and grind bridges and airport runways. Diamond-studded cutters groove runways to prevent hydroplaning and skidding in wet weather.

American states use diamond-studded cylinders to groove and texture roads trying out five different groove patterns to improve cornering and stopping distances in wet weather. Diamond drills bored through a 5-foot reinforced concrete wall 20 feet below the surface of Lake Michigan. Diamond saws cut 18 openings 17 feet by 21 feet through the walls of buildings at the Grand Coulee Dam, and in Philadelphia new trees were planted in sidewalks through holes cut in freezing midwinter with diamond-bladed saws.

In the greatest jewellery and for the world's most precious stones, diamond cuts and polishes diamond. Every year the world produces about 7 million carats of gem-quality diamonds, and half the world's total comes from southern Africa. Demand for gemstones increases and South African revenue for their sales almost doubled in the 1960's.

Gemstones flash in the world's headlines as at Christie's famous jewellery sale in Geneva in May 1969. Here magnificent jewels from Queen Marie-José of Italy, another European Royal House, and the late Nina Dyer, and several other distinguished persons were sold in the Hotel Richemond. A descendant of Mrs. Joseph W. Drexel of Philadelphia and New York sold a diamond necklace of 50 stones weighing a total 103 carats for £21,413. Nina Dyer's famous 'Panther' jewels made by Cartier of Paris sold for £19,749, her black pearl and diamond ear clips fetched £29,187, and her emerald and diamond clip brooch (also by Cartier) made £87,161. Then came her two magnificent rings made by Harry Winston of New York.

Winston started on his own in New York in 1919 with £2,000. He bought jewels from deceased estates, circulating likely vendors from the 'Social Register' and with the help of attorneys who tipped him off when estates were about to be probated. In this way he bought the 'Hope Diamond' and the 'Star of the East'. He had a favourite 152 carat stone from Jagersfontein, and the Jonker as a 726 carat rough. He owned the 426 carat rough from The Premier Mine from which he had cut the 128 carat 'Niarchos'. He could, it was said, spread $12 million worth of gems across his desk from his private safe.

The first of Nina Dyer's stones made by him, a superb navette weighing 27¼ carats fetched £107,019 at the Geneva sale. The other, also made by him, but this time a superb emerald-cut diamond weighing 32⅛ carats, was sold for £114,947. And there are also of course, such diamond rings as the magnificent ones worn by Miss Elizabeth Taylor.

Geneva and the European and American millionaires at the jewellery sales were a far, rich cry from the bleak ridge out by the Finsch Mine. So too is London where in January 1970 De Beers Central Selling Organization reported record sales of gem and industrial diamonds for the second year running. Only once in the sixties has there been a pause in annual growth. The 1969 figures hit a new peak of £288,538,255 worth of sales.

Out at Finsch, there are still a few solitary diggers grubbing away in the earth. In sight and sound of the million pounds of De Beers machinery recovering glittering shoals of diamonds, a few independent prospectors are still hopefully at work. What the old men did in the early days, what Barnato, Rhodes and Ernest Oppenheimer did, and what De Beers are doing now, still lures them on. After all, it is no time at all since Mr. Allister Thornton Fincham was exploring on this very farm Brits for some asbestos.

Just beyond Finsch's brand-new town of supermarkets, banks, and a watered golf-course all laid out and occupied in months, a fissure of kimberlite spreads along the ridge. It is only 3 foot wide. This little trickle of hopefulness springs out sideways from the world's richest diamondiferous pipe. The fissure's soil contains a few stones, maybe more. The prospectors dig trenches into it with spades, and sort and sift. There is no water on the spot. Finsch Mine brings its own in pipelines 50 miles from the Vaal River. Crouched over their holes along this vein, living meanly under canvas and just above the breadline, the handful of diggers beg water for their cans and pray to find one little diamond a month to give them food to live.

Next to the roar of the mighty mine they toil away as hopefully as those first prospectors did in Kimberley a century ago. There is something magic about diamonds. They do not lightly let you go.

Bibliography

BATEMAN; A. M. *Economic Mineral Deposits*, New York, 1950.
BRYDEN; H. A. *A History of South Africa*, 1904.
BURGESS; P. H. E. *Diamonds Unlimited*, 1960.
CARSTENS; J. *A Fortune Through My Fingers*, 1963.
CHILVERS; Hedley A. *The Story of De Beers*, 1939.
CROOKES; Sir W. *Diamonds*, 1909.
DOUGHTY; O. *Early Diamond Days*, 1963.
Encyclopaedia Britannica, eleventh edition.
EVANS; J. *History of Jewellery*, 1953.
FITZPATRICK; James *The Transvaal from Within*, 1900. *South African Memories*, 1932.
FLEMING; I. *The Diamond Smugglers*, 1960.
FORT; G. Seymour *Dr. Jameson*, 1908. *Alfred Beit*, 1932.
FULLER; Sir Thomas E. *The Right Hon. Cecil J. Rhodes*, 1910.
GARRETT; F. E. and EDWARDS; E. J. *Story of an African Crisis: The Raid*, 1897.
GREEN; Lawrence G. *Like Diamond Blazing*, 1967.
GREGORY; Sir T. *Ernest Oppenheimer and the Economic Development of Southern Africa*, 1962.
HAHN; Emily *Diamond*, 1956.
HARRIS; Sir David *Pioneer, Soldier and Politician*, 1931.
HENSMAN; H. *Cecil Rhodes: A Study of a Career*, 1901.
HERBERT-SMITH; G. F. *Gemstones*, 1958.
HOLE; Hugh Marshall *The Jameson Raid*, 1930.
IDRIESS; Ian L. *Stone of Destiny*, 1953.
JEFFRIES; D. *A Treatise on Diamonds and Pearls*, 1750.
JOURDAN; Philip *Cecil Rhodes: His Private Life*, 1911.
LEWINSOHN; R. *Barney Barnato*, 1937.
LE SUEUR; Gordon *Cecil Rhodes*, 1913.

LOCKHART; John G. *Cecil Rhodes*, 1933.
MACDONALD;; James G. *Rhodes: A Life*, 1927.
MCCARTHY; J. R. *Fire in the Earth*, 1946.
MICHELL; Sir Lewis L. *The Life of the Right Hon. C. J. Rhodes*, 2 vols, 1910.
MONNICKENDAM; A. *The Magic of Diamonds*, 1955.
NORWOOD; V. G. C. *A Handful of Diamonds*, 1960.
PAYTON; Chas. A. *The Diamond Diggings of South Africa*, 1872.
DU PLESSIS; J. H. *Diamonds are Dangerous*, 1961.
RAYMOND; Harry *B. I. Barnato. A Memoir*, 1897.
ROSENTHAL; Eric *Here are Diamonds*, 1950.
SCULLY; William Charles *Reminiscences of a South African Pioneer*, 1913.
State of South Africa Year Book, 1969.
STATHAM; Reginald *Paul Kruger and his Times*, 1898.
STREETER; E. W. *A Short History of Diamond Cutting*, 1888.
Precious Stones and Gems, 1898.
The Great Diamonds of the World, 1882.
SUTTON; J. R. *Diamond*, 1928.
TAVERNIER; J. B. *Six Travels*, 1684.
TAYLOR; William P. *African Treasures. Sixty Years among Diamonds and Gold*, 1932.
TOLANSKY; S. *The History and Use of Diamond*, 1962.
WEINTHAL; Leo *Memories, Mines and Millions*, being the Life of Sir Joseph B. Robinson.
WILLIAMS; A. F. *Genesis of The Diamond*, 1952.
WILLIAMS; Gardner, F. *The Diamond Mines of South Africa*, 1902.
WILSON; R. F. *Cecil J. Rhodes*, 1900.
YOUNGHUSBAND; Francis *South Africa of To-day*, 1898.

Index